Legacies

Library of Congress Cataloging-in-Publication Data

p. cm.

Includes index.

ISBN 0-7803-9996-X : $20.00

1. Electric engineers--United States--Biography. I. Campbell, Mary K. (Mary Keren), 1956- . II. Institute of Electrical and Electronics Engineers. Life Member Committee.

TK139.L44 1994

621.3'092'2--dc20

[B]

94-42255

CIP

JP27023

1. J. Rennie Whitehead
2. Weislaw Barwicz
3. Leo Berberich
4. Marc Chauvierre
5. Aubrey G. Caplan
6. Julius Stratton
7. E. E. Thompson
8. Jerome Kurshan
9. Nathan Gordon

10. Robert Newman
11. Francis J. Heyden
12. Kenneth Miller
13. James Braxdale
14. Gustave Bliesner
15. John DiNucci
16. Marc Chauvierre
17. Edgar C. Gentle, Jr.
18. William H. J. Kitchen
19. R.H. Eberstadt
20. John R. Wine
21. Simpson Linke
22. Herbert Butler
23. Hubert H. Humphrey

Legacies

Edited by
Mary K. Campbell

Institute of Electrical and Electronic Engineers, Inc.
Life Members Committee

Acknowledgements

We'd like to thank the many people who contributed their time and effort to this project. Legacies couldn't have happened without them:

Michael Bramwell
Mary K. Campbell
Suzanne L. DeFilippo
Robert F. Lawrence
Gail E. Leeman
Diana Pladdys
Deborah Rosenberg
Robert Sacks
Helen Shiminski
Deanna Watrous

And they said it couldn't be done!

Contents

Foreword

A book about engineers, "Legacies," has a nice ring to it doesn't it? How did this book begin? Looking back on my own career, around '43, I was a senior graduating from college and commencing my career in electrical engineering. Early on, in fact in my senior year, I became active in the AIEE, American Institute of Electrical Engineers, which later merged with IRE into the IEEE (1963). I soon discovered that publications of technical work were, and I think they still are, the basic foundation of the professional society. The benefits and impact on the technological community and society as a whole, were and are truly significant.

The benefits to the author were significant, too. The author gained recognition among peers and in the technological community. However, not every engineer had the opportunity, nor maybe the inclination to write and publish. Even authors who have written about their technical work may not have revealed much about their careers.

And so, at some point in time as a new member of the Life Members' Committee, it struck me we should extend an invitation to Life members to write about their careers. Publication in an edited form would provide entertaining reading to other Life members; and the results could be stimulating to young potential engineers. And what better subject could an engineer write about than his or her accomplishments? Armed with the concept of biographical sketches and the support of Staff and the other Committee members, we decided to proceed with our experiment.

In the *Life Member Fund Newsletter*, we asked any interested Life member, to submit his or her "Life story" for review, editing and publishing. A broad outline was provided for general guidance although the writer was encouraged to write as much as he or she felt like. We did not want to inhibit content or length of the submittal, although we did mention the necessity for final editing. We wanted to produce entertaining reading based on insights of the authors' careers, their first jobs, bosses, accomplishments, why be engineers, humorous life interest experiences, and so forth. The response was excellent, well beyond our expectations—some 200 replies. As the stories poured in, it was a bit overwhelming because the LM Staff was committed to the project on a spare time basis.

I think, however, there is a little more to say about the underlying reasons for this project. There is a certain curiosity about an engineer—what does an engineer do—why be an engineer—how does one become an engineer? I have two nieces who for years thought I drove a train (a secret, unfulfilled ambition by the way).

I must admit to an introspection or two on my own "curiosity" about engineers. In fact, in April of 1983, as an IEEE Board Director, I spoke at the Pittsburgh Sections Recognition dinner. The subject was "The ABC's of EE." I talked about an engineer's *ability* to wonder, be *adaptable*, look for ways to make products *better*, and our *commitment* to our work. Once you get an engineer going on a project, it's pretty damn hard to get him or her off it.

I'm not the only one to evince a curiosity about what makes a person an engineer. Robert W. Lucky, whose column "Reflections" appears in *IEEE SPECTRUM*, addressed this issue in an editorial (July, 1991). Here are a few quotes out of context that seem appropriate here.

"People often ask me how I became an engineer, as if to say, how did you pick up this terrible disease?......Usually I'm bored with this topic, so when someone asks that question, I simply say that I was born an engineer. That cuts the conversation short, with the questioner nodding in apparent understanding....'Of course....What a terrible burden to carry through his life. Amazing that he turned out so well—considering.'.....Amused by this fantasy, I polled some of my engineer friends to find out how they became engineers. 'Do you feel that you were born an engineer, or made into one?' I [Lucky] *asked. No one admitted to being born an engineer. Instead they give almost identical stories. 'Well, I always liked to play with radios, (or some such mechanical or electronic gadget),' they would say. Then they invariably finished with a note of pride. 'I was always good in math and science,' they said to explain why."*

Then Mr. Lucky introspectively questions a whispered heresy that "*perhaps engineering is a trainable skill—that someone who hadn't been born an engineer—someone who hadn't played with mechanical and electronic things or hadn't necessarily aced the math and science courses—that such a person could actually be admitted to our sacred profession?*"

Returning to my own thoughts, I recall an interview for my first job. The interviewer, from Westinghouse, asked very pointedly, "How did you decide you wanted to be an engineer?" I suppose I felt defensive, I could have admitted to the revealing engineer gene trait, "I played with an Erector® set as a youngster." But, in all honesty, I don't remember exactly how I did answer the question. It must not have been too damaging because I was hired!

I also recall feeling that the organized thought processes of engineering were the epitome of technological intellect akin to the precise logic of mathematics, and well above the status of some other disciplines, e.g. accounting, economics and financial management. The last of these was a curse to endure and overcome to achieve an engineering diploma. Later on, reality proved the "real world and market forces" would impact significantly on my idealized engineering solutions to problems.

Since this project began, I've also been saving a response to an editorial in the Pittsburgh Post Gazette, a local Pittsburgh newspaper. I think the letter is so good, I've been keeping it for this very occasion. It was written by an engineer, Dr. Joseph M. Newcomer of Pittsburgh.

But first a bit more background. The editorial, in paraphrase, stated that, "All he ever managed to build from his Erector® set was junk, nothing he built ever looked like those neat pictures in the accompanying manual, and the Lincoln Logs was a pointless toy because there were never enough pieces to build anything."

Dr. Newcomer responded with this letter to the editor. It was published November 27, 1991, and is quoted verbatim as follows:

"Years ago, I had a Tinker Toy set. I never built the windmill or any of the elaborate things shown in the book that came with it. But I persevered. I also had an Erector set. I never built the parachute jump or the Ferris wheel—I only had the small set anyway. And I must admit that the small Lincoln Logs set never was suitable for more than a simple log cabin.

But there was very little point in building the things that were in the books. They already

But there was very little point in building the things that were in the books. They already existed. The fact that they could be built was not terribly interesting. I wanted to build things that hadn't been built before.

And somehow, the junk I built amused me. True, it wasn't as slick as the stuff in the books, but it was my junk, and it hadn't ever been built before, by anyone. It exercised my imagination. I discovered that I liked to build things that hadn't been built before. I built a flight simulator out of Tinker Toys, sometime around the age of 10, which had an airplane mounted with strings. Getting the stick to move right was a challenge, and I never quite got the linkages all to work, but it was a lot of fun.

Somehow, this interest in building things continued. The creativity my 'junk' projects evoked turned into a doctorate in computer science and a quite successful career. I am very happy doing what I am doing. I now have the world's largest Tinker Toy set and the world's largest Erector® set. Except that instead of knobs, sticks, girders and bolts, I have RAM, hard disks, and other such goodies. I wish I could have had this thirty-odd years ago!

So don't denigrate those 'junk' projects. You never know where they are going to lead.

I think Dr. Newcomer's letter adds a nice touch in putting a perspective on LEGACIES." Maybe it suggests today's computers will be the forerunner of something we can't even imagine 30 years from now. Maybe the video game kids of today will be engineers, not "born" but like latent film images "developed" to extend technology into the future. Who knows, maybe "LEGACIES" will be a stimulus or catalyst; better yet maybe you will play that role.

Many thanks to all of those who have participated; and good wishes and good reading to all of those who do so.

Bob Lawrence, Chair
1991-1993 Life Members Committee

Chapter one

Childhood tales

John L. Shaw

My interest in engineering began by osmosis. I cannot remember any time it was not a "given" that I was going to be an engineer. We got our first television set in 1936 when I was in my early teens. Yes, I said 1936! My father was Dr. G. R. Shaw (Fellow IEEE). At the time, he was Chief Engineer of RCA's Tube Division. So for the next several years, we had a succession of the latest TV sets to evaluate. Our first one had a twelve inch green (Wolframite) tube. The set itself was designed to look like a piece of furniture with no screen or knobs visible so no one would ask questions. The tube had to be mounted vertically because it used electrostatic deflection. The lid of the cabinet (containing a front surface mirror) was opened to forty-five degrees to view the tube and access the fourteen knobs.

RCA engineers had installed an antenna on top of the Empire State Building for W2XBS. I remember Bill Hickock telling of climbing a ladder to adjust the antenna and having to break an inch of ice from each rung as he went up.

(Bill Hickock got me started building radios by patient advice and generous donations from his personal store of radio parts. I astounded my high school classmates when I built a radio inside a book cover using the "peanut" tubes [1R5, 1T4, 1S4 and 1S5] while they were still in development.)

Broadcasts began at eleven p.m. (for security reasons) and ran to one a.m. So we went to bed early and got up for those two hours to view (mainly) a "Felix the Cat" cartoon and newsreels of the 1936 Florida hurricane. My prime duty was to record horizontal and vertical resolutions from the test pattern. Exact dates are hazy but soon NBC was broadcasting during weekends and we were visited by many RCA engineers working on the development. Names I recall are Dee Wamsley, H. B. Law, Stan Umbreit, Ed Herold, Wally James, Otto Schade, A. D. Power, Tony Lederer and, of course, Vladimir Zworykin.

I learned to never say never when Otto Schade told us he had picked up an English TV station the day before. The experts around our TV agreed that it must be due to an unusual Heaviside layer condition and that regular transmissions from Europe would, of course, be impossible.

About 1939, the TV secrecy wraps came off, programs were better and were shown during the day. At first the visitors who came to see it were Dad's business friends so I got to meet people including Allan Dumont, his chief engineer, Tom Goldsmith and Claude Shannon of Bell Labs. Also the neighbors began to drop in for a look. One weekend I tallied fifty visitors. When I left for the University of Wisconsin in 1941, Dad had one of two projection sets (Sarnoff had the other) which used a five inch TV tube pointing downwards into a Schmidt optical system. It was actually quite good.

Cyril G. Veinott

Although I was born in 1905 in Somerville, Massachusetts, my childhood was spent in southern Vermont. For the first eight years, it was in the small hamlet of West Townshend, and then some ten years more in Chester, a town of fifteen hundred. In summer, travel was literally by horse and buggy. In winter, the horse drew a sleigh because the dirt roads were never ploughed—the snow was just packed down by huge horse-drawn snow rollers.

Our telephone was mounted on the wall, with several parties on the same line. We could call anyone on our line by hand-cranking the appropriate number of rings. Anyone on our line could listen in, and even interrupt if they so desired. For "long distance" calls, we had to ring the operator and wait interminable periods of time to reach the called party.

Homes generally were not wired for electricity. For the bright light needed for reading or studying, we had a kerosene lamp that used a fragile mantle, just like the ones commonly used on gas lamps. Generally, we did not have inside plumbing; the one (or two) holer would often be in the corridor that joined the house to the attached barn. When one went to the bathroom in wintertime, one did it as quickly as possible.

Our first car was a Ford Model T. To start it, we used a hand crank. An electric starter and demountable rims were extras. Side curtains didn't keep out all the rain, nor did they keep one warm in cold weather. Headlights were supplied from the ignition magneto mounted on the engine flywheel. When the engine slowed down on a hill, the headlights would go almost out; often the driver had to drop into second gear so the engine would run faster and the lights got brighter.

Alfred W. Barber

In 1920, I (at age fourteen) bought my first radio magazines and began to hanker for a vacuum tube to experiment with. *The Radio News* (August, 1920 on page 68) carried a full page ad for the Audiotron, "the first and only amateur vacuum tube licensed under Fleming. Vacuum tube patents are basic and have been sustained by the Federal Courts."

This tube sold for six dollars. My father said it was too much to spend on a hobby but I said I wanted it; I had the money and I'd never ask to buy another thing in the radio field. (I broke that promise thousands of times over.) I rode my bicycle thirty miles to buy my Audiotron.

I built a fine double tuned regenerative detector radio in my high school woodworking class and had probably the best single tube radio around. While there were several hams in the high school and I was a member of the Radio Club, I never became a ham or got an amateur license.

Kenneth G. McKay

As an only child reared in comfortable circumstances, I had many opportunities to develop solitary pursuits. At age five, I became enamored with a Meccano set—a British construction set somewhat like the Erector® set. Meccano sets came in different sizes and different complexities. An ingenious marketing tactic provided auxiliary sets to enable one to progress upward.

For seven years, I saved most of my allowance for upgrades until I was building a working grandfather clock, a motor chassis with a differential and three speed gear shift, and bridges that spanned the family den and adjacent hallway. The bridges brought only muffled complaints as my parents' living space was progressively restricted. In retrospect, they were remarkably restrained.

Emil Gaynor

I honestly cannot remember when I first thought about becoming an engineer. It must have been when I was about ten. I tried putting together a telegraph set from nails, scrounged wire, some pieces of a tin can and an old 1.5 volt dry cell (# 6). I started checking out electrical texts from the library even though I couldn't understand them. This all took place in the Big Apple (New York City) where we were all severely impoverished by the depression.

The junior high school had some excellent shop courses that added to the urge to be an engineer. I was extemely fortunate in being able to go to the Brooklyn Children's Museum—an hour's subway ride from the Bronx where I lived. This outstanding facility was a hands-on place. Here we learned about and handled different kinds of reptiles, worked with microscopes, tested many minerals for hardness and their specific gravity, along with studying insects and birds. This was a very important step in learning how to learn.

Enrico Levi

I was born near the end of World War I in Milano, Italy. Most probably, my family had lived in Italy for more than two thousand years, and at various times, took active roles in creating its history. It was not until the promulgation of the racial laws in 1938 that I discovered that I was not an Italian, but a Jew.

I decided to become an engineer at the age of four when I received a steel construction toy called "Meccano." I never regretted making that decision.

Frank T. Luff

My first thoughts about becoming an engineer probably came in 1937 when I was a sophomore in high school in the small farming town of Palmyra, Nebraska. I had built a crystal radio set using the procedures and methods of one of my high school mates. It didn't work. That made me very determined to try again using more reliable methods. I found a book (I think it was called *Radio Engineering*) that gave step-by-step procedures about how to build a very complicated and fancy looking crystal set. It worked.

I heard the Orson Wells invasion by the Martians on that crystal set. I was quite concerned about it; however, we lived far out in the country, and places like St. Louis and Chicago seemed worlds away to me.

I attended a one room country school house that included grades one through eight and had one teacher. Eighth grade students had to pass a series of tests before they could go on to the high school in town. I hadn't been a very good student and my parents weren't expecting me to pass the tests. But the teacher (the only one for the eight grades) worked with me and I liked her. I not only passed the tests, but I made the county honor roll! The county school teacher, Miss Fender, made a difference in my life.

The high school was geared to teach what farm boys and girls needed to know. Farming was done almost completely by horse drawn machinery in those days. Only simple algebra and general science was taught. I had no high school classes in geometry, trigonometry or physics or chemistry. No foreign language. The country was in a deep depression. My father was forced off the farm and had to work on WPA (Works Progress Administration). There was no hope of going to college; no hope of getting any kind of job when high school was over in 1939. I spent a year in a 666 Camp then joined the Marine Corps when I was old enough. A year later we were at war with Japan.

Harold W. Lord

(1905-1922) My interest in things electrical began in 1915 when I visited, with my father, the Panama Pacific Exposition in San Francisco. This was held to celebrate the opening of the Panama Canal, which meant so much to the west coast of our country.

While in San Francisco, we visited the toy department of a large store, probably the Emporium. There I saw and was very interested in a toy model of an electric motor which had a large eye bolt for lifting it! My Dad saw my interest in it, and it was one of my presents the following Christmas.

In those days, the telephones were operated by No. 6 dry cells. When no longer useful, the batteries were discarded on the trash heap in back of the central office. I collected some of them from the trash heap and rejuvenated them to some extent by punching holes in the zinc enclosure and immersing them in cans of salt water. The cells would make the motor run, but would not carry much of a mechanical load.

My present the following birthday, when I was eleven years old, was an Amertran transformer. It had a tap-changing switch by which the voltage could be varied from about two volts to twelve volts. It operated my motor just fine as it was a series-type motor.

I had not had my transformer very long before I pried open the lid very carefully so I could see the core and coils. It was a two-legged core with the primary on one leg and the secondary on the other leg. The leakage reactance of this coil arrangement was sufficient to limit the short-circuit currents to safe values. The lid could be snapped back in place, so satisfying my curiosity did no damage to the transformer.

The Klaxon Horn, used by many automobiles in those days, was driven by a six-volt electric motor. I do not remember how, but I acquired a discarded Klaxon Horn and salvaged the motor. I rewound the motor armature to operate at a higher voltage, and it had more torque than my toy motor.

The Model T Ford automobile used a vibrator type spark coil for each of its four cylinders. When these coils failed to deliver rated voltage they were discarded. I acquired one of them and it operated fine from my toy transformer. It provided a mild electrical shock if a piece of metal held in the hand was brought near enough to the terminal of the spark coil so that a spark would jump from it to the metal.

Using the transformer, I devised a "shocking" electrical game. One terminal of the transformer was connected to a tin can of about four inches diameter and six inches high and three quarters filled with water. The other terminal was connected to a piece of metal, about an inch wide and eight inches long. With one wetted hand pressed on the metal bar, the object was to draw a coin from the bottom of the can at the highest voltage possible. I managed to do eight volts, but many couldn't even do six volts.

Leo Berberich

In 1918, at twelve I watched with excitement as the workmen unloaded and installed the DELCO generator set at the family farm. I didn't know much about electricity, but I knew our family would have running water in the house and even a flush toilet. I was sure I would not miss trips to the privy.

The gasoline powered 32 volt generator and its set of 16 batteries would save me from studying by the light of a kerosene lamp. I was excited about the installation of the electric system. I could not resist watching the workmen string every foot of the wiring in all the farm buildings—even the chicken house. I wanted to know a lot more about this mysterious electricity. I wanted to learn how to manipulate it.

I was born on November 19, 1906, the son of German immigrant parents, on a small farm north of Petersburg, Virginia. I graduated from Saint Joseph's High School in Petersburg in 1924. In the fall of that same year, I entered the Johns Hopkins University in Baltimore, Maryland.

My first real challenge was defending my decision to attend college. THE concept of college was not warmly received by my father, Moritz, who expected me, as his eldest son, to continue on with the family farm. My father had already come to rely on me to do important farm chores such as plowing the fields, cultivating and harvesting the crops, making hay and milking the cows. However, my favorite chores involved anything which required me to drive the family's 1917 Model T Ford. Hauling the farm's milk to the local dairy or taking my sisters and brother to school in the Ford (acquired at the cost of three hundred and sixty-five dollars) pleased me enormously.

Kenneth Sturley

It was in the 1920s that my lifelong enthusiasm for radio was kindled. As a schoolboy on holiday in Aberdovey, Wales, I was thrilled when into the small port came Marconi's yacht, *Elettra*, on a visit to the transmitting station at Towyn. I was able to go on board to see the radio cabin and the great man himself. Naturally I made my own crystal set and later a valve receiver to listen to broadcast programs.

Philip Morris

When I was around six, I lived in Jackson Heights, New York, a few blocks away from Holmes Airport. (Holmes Airport was closed when Laguardia Field was opened.) Also, it was a pleasant walk to the North Shore where the North Beach Seaplane Base was located. North Beach was the base that was rebuilt into Laguardia Field. I visited both of these places quite frequently. Not too surprisingly, I wanted to become an airplane pilot.

Unfortunately, it was about this time that it was discovered that I needed to wear glasses. The eye doctor said that there was a possibility that I might outgrow the need to wear glasses. For many years I lived with the hope that this would happen so I could become a pilot. Gradually, I realized that this was not going to happen. Going for what I considered the next best thing, I decided that I wanted to be an aeronautical engineer. If I couldn't fly planes, at least, I could design and build them.

Granville Porter

I was born Granville Dwight Porter in Ne, Ohio in 1916. In 1919, after a twenty-five year marriage with my father, my mother opted for life as a single parent in California. I went to first grade at Los Feliz School at Hollywood Boulevard and Vermont Avenue. After graduating from Polytechnic High in 1934 with a pre-engineering major, I attended LA City College during 1934-35. The following year, I attended the Radio Institute of California where my instructors included Mr. Sidney Bertram and Dr. Fred Stahl.

(My mother had been successful in business, providing me with the proverbial "silver spoon"; however, she lost it all in the 1929 crash. I first earned our livelihood by working in a small electric shop on Pico Boulevard for three dollars a week and later for five dollars!)

Garabed Hovhanesian

To understand where my "engineering" originated, one must understand where I originated. Both my parents were born before the turn of the century, approximately 1890, in a village in eastern Turkey. This had been the homeland of the Armenian people for three thousand years. My parents were married in 1907. In 1909, my father fled Turkey to avoid being conscripted as cannon fodder for the Ottoman Turkish Army. Shortly after he fled to America, the Turks or their surrogates beheaded both my mother's and father's parents, around 1910.

I have no record other than stories told by my mother who was a witness. She suffered the terrible massacres and marches of attrition through the Arabian deserts of Syria and Iraq from 1910 to 1918, the end of World War I. While she survived by her wit and some luck, thousands of others were not so fortunate. Their bodies were dumped like so much garbage into the Euphrates and Tigres Rivers if, indeed, they survived the march to those rivers.

When the Allies finally beat the central powers including the Turks, my mother was liberated by the British Eastern Army. She made a circuitous route to the United States. She landed in New York City at Ellis Island on July 3, 1920, and saw the fireworks in the harbor. On April 28, 1921, I was born in the city of Lowell, Massachusetts. Everyone worked in the textile mills but there was still no peace.

The textile industry was already dying in the Northeast, especially in Lowell. All the mills were moving to the Sunbelt. Lack of skills, language problems, and the harsh realities of ethnic discrimination and economic privation forced my parents to move to Worcester, Massachusetts. Shortly thereafter the crash of 1929 occurred. This was probably the final blow for my father. He died in 1936 at the age of forty-six. He died of meningitis but his heart was already broken. How was he to raise two sons and feed a family?

Before my father had died, he had already inculcated me with his basic rule for survival, "nobody can take away from you what you have between your fingers and ears." His wish for me was to go to trade school. By the age of nine, I already knew what a patent was and what its value could be. To me, it meant freedom from want.

I was an avid reader and my first hero was Thomas Alva Edison, my next was Henry

Ford. There were many fictional heroes but another author, Sir Guy Eddington, caught my imagination. I will never forget his "popular" explanation of Einstein's theory of relativity.

Paul Burk

As many other Life Members , I got into radio about 1929 via a catwhisker, a galena crystal, a pair of headphones and about fifty feet of antenna wire. The livingroom, bathroom and bedroom were wired for headphones from the single crystal sitting on the livingroom window sill. Subsequently, I advanced into vacuum tube radio by acquiring three discarded sets at the old Cortland Street (New York City) flea market.

At one time, I had a collection of 00A's, OlA's, 71A's, 99's, Kellogg bulb top connected heaters and other vintage items of that era. Typical of many youngsters of that time, I was into model planes and photography as well as radio. But since radio repairing was the only self supporting hobby, I stayed with it and have been with it ever since.

Leslie Balter

I never remember consciously deciding to become an engineer. My father was a mechanic who had attended a trade school. He had the ability to take anything apart and put it together again working better than it did before. I assume, I inherited some of his abilities.

The first thrill of radio I can remember was gathering around the old battery-operated 01A tube set with its monster variable ganged condenser (capacitor) in New York City listening to WXYZ, Niagara Falls, New York. That was a thrill (1928, I think). I knew I was destined to be in electronics when the following day I twirled the knob and burned out three tubes by raising the battery filament voltage excessively high. It used a storage battery for filament power with a series rheostat to limit voltage.

In the Boy Scouts, for the Radio Merit Badge, I learned the pleasure of winding enamel antenna wire around an oatmeal box coil form, constructing a slider by scraping enamel, and hooking up galena crystal with a catwhisker to hear local radio stations. I think, from that point on, I was hooked. I received a short wave receiver for my thirteenth birthday, learned the code, and at age fourteen, passed the code test and received Amateur Radio License W2HRT (a call I hold to this day). At that point in time, I believe I was one of the youngest, if not the youngest, "hams," licensed in the Second District in 1934.

Edgar C. Gentle, Jr.

I became an engineer because I just liked the idea of designing something. My father was with AT&T (an engineer by the working experience and training he had there). As I was growing up, I spent a lot of time with him and other people who had engineering or technical backgrounds.

William A. Edson

My brother Jim, who was seven years older, was already studying electrical engineering when I entered high school. He had a job at Bell Labs before I graduated. Also, my Uncle Jim was a professional engineer, and my father had some training in engineering. So it was almost preordained that I should go into electrical engineering.

Theodore Schroeder

Various influences during my youth resulted in various ideas as to "what I was going to be" but they always involved some form of engineering. Architecture was it at one point because of a fascination with building construction and architectural drafting; then mechanical engineering because of mechanical drafting (in high school); finally I decided electrical engineering because of the idealistic thought that supplying electricity to the public must be the greatest contribution to society.

Henry F. Seels

I, Henry F. Seels, started my aviation interest at age four in England. My parents and I, in 1914, fled from Antwerp, Belgium to London—where during WWI—we watched German planes blow up English observation balloons over London. Then in 1916, we shipped over to New York City via Ellis Island. My parents paid a ten dollars a head tax fee to enter the United States of America.

My early youth was spent between Hell's Kitchen and the Village as I watched the house we maintained being changed from gas service to dc electric. The dc was changed to ac and many electrical appliances were exchanged by the Edison Electric Company. These early days exposed me to many service problems.

Nigel Stace

For reasons best known to them, my parents early on decided that I should become a civil engineer designing bridges, roads, and perhaps dams. So much so, in fact, that Meccano's "Engineering for Boys ages 7 to 70" advertising slogan caused my parents to buy me a No. 1 Meccano set and then annually provided supplementary sets up to No. 6A, the largest one made.

Their scheme worked well except for one aspect. Although my constructional process flourished to the extent that I won a major prize in a national competition, my interests were in mechanical, working models rather than in static, structural ones. For working models, some driving mechanism was needed. Eventually, I acquired a robust Meccano clockwork motor. But its restricted running time and regular rewinding became a bore. When an electric motor was announced by the New Zealand agents, it became my driving ambition to have one. At last I did—but only for one memorable week!

But what a week that was, with almost perpetual motion plus intermittent electric shocks. As I remember it, the motor cord was simply plugged into the nearest lamp socket. (The only plug point was in the kitchen and that was reserved for ironing.) However, the English-made motor and its insulation were designed to operate on half the New Zealand's main voltage! An electrician (wireman in those days) came to my house and declared my motor hazardous and it was taken away.

Samuel Sensiper

My first recollection of an encounter with radio was when I about four or five years old. My father had brought home a device which was reputed to be able pick out voices or music from the air! (Remember, this was about 1923 or 1924.) After some unsuccessful attempts, my father, mother, and sister went to eat the evening meal and I was left behind. I duplicated what my father had been doing and to my amazement and theirs, when they responded to my shouts, I had tuned in to a broadcast. I'm quite sure this early encounter with radio had no impact on my later career choice.

However I'm sure that my next major encounter was influential. It was an advertisement in a long since defunct magazine, *American Boy*, as I recall, regarding the ARRL (American Radio Relay League). The advertisement extolled the fun of being a radio amateur. I sent for the information which led eventually to my building a tube receiver and transmitter with a Zepp antenna, to acquiring the license W2GXE, to belonging to the local amateur radio club, etc. As I recall, I acquired my license in about 1933, and operated actively until 1935-1936.

Walter Schweiss

I was fascinated by hearing music from my uncle's headset connected to a crystal radio. Owning few toys, it was only natural to play with mother's sewing paraphernalia. These items were wooden spools, yarn, string, rope, rubber bands, and popsicle sticks. Later, rusty nails and fruit boxes/baskets provided the materials to build handmade go-carts, roller scooters, wooden toys, playhouses, and so forth.

School also exposed me to manual trade classes and music training that created many interests. Hobbies followed with some zeal in old battery radios gathered from neighbors' attics.

Reading newspaper articles about building one's own short wave radio receiver intrigued me immensely. These were the first instructions I gained in the understanding of electronic components. As a radio enthusiast, and becoming a member of the American Radio Relay League (ARRL), a radio ham station evolved very quickly.

Serious interests developed during high school by way of my electrical shop training from Mr. Chambers and electrical machinery training from Mr. Thomas. They both gave me extra assignments to instill further interest in electrical theory. Dr. Greenwood, the mathematics instructor, became a great inspiration in my learning at an early age. Technical knowledge was also gained through the Olney Amateur Radio Club, which my friends and myself organized.

John R. Smith

The desire to become an engineer occurred one Sunday at a church service while I was still in grade school. The preacher was making a point for education. He told the story about a man who was having trouble with the diesel engine driving a generator. The engine company sent a mechanical engineer out to fix it. In a few minutes, the engineer adjusted a couple of screws on the carburetor and the engine began to run perfectly. When the engineer presented the bill—fifty dollars—the man said, "It is well worth the fifty dollars, but how did you arrive at that figure?" "Oh," the engineer said, "ten dollars for turning the screws, and forty dollars for knowing which screws to turn." That story convinced me that I wanted to be an engineer.

J. Rennie Whitehead

I was born in a small village in Lancashire in the northwest of England. By the mid-1920s, when I was seven or eight years old, I was building crystal sets to listen to the BBC transmissions from 2LO, Manchester on the medium wave band and the powerful 200kc/s transmitter at Daventry, as well as commercial radio from Fecamp and Luxembourg, on the long wave band. The crystal was usually a piece of germanium found by sorting through hundreds of lumps of coal outside the coalhouse until the characteristic glint of a germanium crystal could be seen in one. It was extracted and mounted in a chuck with a spiral copper wire (catwhisker) as the contact.

M. E. Scoville

Born in 1905, I grew up as a poor farm boy. Several different farms we lived on had no electricity, no running water in the house and no refrigeration for food. After 1916, we drove a 1914 Model T Ford. Cooking was done on a kitchen range burning corn cobs, wood or ear corn when corn prices were low. Vegetables from a large garden, fifty bushels or more of potatoes and possibly a few bushels of apples were stored in a cave for winter use. We milked four to ten cows from which some cream was sold. Chickens provided some eggs for sale. Cream and egg sales resulted in money for very limited groceries bought to feed seven children.

Farming and living know how was learned directly from parents or neighbors. My father and mother finished eighth grade country schools. Mother taught school for two years before getting married. Nearly all our farm crops of corn and forage were fed to our farm livestock consisting of horses, cattle, pigs and chickens. All farm work was done with horses.

My father, being mechanically inclined, owned a corn sheller and power wood saws which were used to do custom work for neighbors. About 1910, the power for these machines was literally an eight-horse circular horse power drive. This worked as long as I sat with a long whip to keep the horses pulling when they heard the saw start on a large log. The "horse" power was replaced in about 1915 by an 8 H.P. single cylinder gasoline engine.

During my last year in high school, I finished a correspondence course on electricity which was more stimulating for me than any of my high school courses. But my experience with electricity was very limited. We had a hand cranked telephone with dry batteries, a Model T with magneto ignition and no other electrical devices. Lightning during thunder storms was always intriguing.

High school teachers urged me to consider going to college. Less than ten percent of high school graduates from Sumner, Nebraska, went on to college in 1924. I had only my own earnings to support college expenses and no fixed plans as to extent of college attendance. After minimal correspondence with three colleges, I selected the University of Nebraska primarily because it was the least costly. When I had enough money, I would start college and continue on as long as I was able to earn money for college expenses.

I worked in a threshing gang and husked corn in the fall of 1924 for neighbors. I earned about one hundred and twenty-five dollars for tuition and room for one semester, starting in January, 1925. Also I ordered a partially assembled OZARKA battery radio, the first in the community, hoping to demonstrate and sell several to neighbors. But reception was unreliable and farmers had no money for radios. The one set was sold at cost.

Alfred J. Siegmeth

I was born on May 10, 1911, in Komarom, Hungary. In 1918, the Austrian-Hungarian Monarchy lost World War I and Hungary's territory was reduced to one third of its original size. After my father returned from his four year military engagement, our family moved to Budapest.

The disintegration of our country was evident. The subsequent economic depression combined with unmanageable food shortages made my early years uneventful and dull. My primitive toys had not satisfied my ambitions. My interest drifted toward the enjoyment of our town's wired broadcast system.

This daily service had been operating since 1893. A Hungarian entrepreneur, Theodore Puskas, had initiated the basic operational capabilities of this "news" network. It provided regular financial, theatrical, local and international reports in addition to newscasts, weather reports, and accurate time announcements.

Basically I loved this information service; but I detested the wire jungles mounted on our roof tops around our city. I hated this ecological eyesore. Around 1922, a school friend of mine informed me, with big excitement, that his father subscribed to a German magazine entitled, *The Austrian Radio Amateur*. I asked him to loan me some older copies of this periodical. This was my first encounter with radio and electronics and I recognized immediately the fact that Marconi's phenomenal "wireless" invention would be extremely useful for dissemination of news and knowledge.

I was captivated when, in 1923, I received with my self-made crystal receiver the "first" experimental radio broadcasts radiated in Budapest. Subsequently, I listened on our wired network to a lecture series about the controversial evolutionary theories of life outlined by Charles Darwin. This complex information together with the mystical complexities of electronics were somewhat above my head. I started to realize during my later adolescent years that it would be to my benefit to acquire a college level knowledge base.

Warren L. Braun

Dad changed churches in 1930 and moved to downstate Illinois to a tiny, sleepy village near Quincy, called Golden. However, "Golden" its promises had been at one time, they had faded to a tumble down gray. Shortly after we arrived, the second bank in town closed. Many businesses were shuttered, and Dad's salary was cut not once but twice.

Feeding the family became a daily

logistical chore. Due to the drought, which endured for three years, there was virtually no garden produce to fill the family larder. It was a farming area. The depression and drought sent pig prices so low that entire litters of pigs were buried because no one could afford to feed them. Corn was burned in stoves in place of wood. Our family, now two boys and two girls, worked in every way we could to help the family survive. The parsonage was old, heated by pot bellied stoves with no heat in the bedrooms. Those were cold and frightening years.

The dust clouds came and went, leaving their signature on every window sill, but still no rain. The summers were hot and dry and the winters biting cold. School was two miles away across the railroad tracks. With holes in the shoes, your feet would get wet, resulting in what were called "chilblains." Cardboard liners helped, but Dad's ingenuity and ten cents got us new glue-on rubber soles. It helped until they let loose with an embarrassing flap-flap. It was pure misery in either summer or winter. That experience left me with a poor appreciation for the mid-west to this day.

Early in my high school years, I developed an interest in radio repair and ran a somewhat profitable repair business under the trade name of "Superior Radio Service." I am sure my proficiency did not live up to the euphemistic name. However, it was profitable, operating out of a henhouse converted to a business by my personal efforts. I guess the aroma helped fertilize the OJT (on the job training). For by the time I finished high school, I had a rather good working knowledge of circuit theory and its practical application. I had even constructed my own multi-meter to do the servicing.

Joseph F. Furlong

It was the horn gap that did it. Whenever the Saturday movie serial had a lab scene, there was this diverging pair of wires with an electrical arc rising up and disappearing into the air. Of course, I didn't know then that this was a horn gap or what its purpose was, but I thought it would be great to work with this kind of thing. So I decided to become an electrical engineer.

William H. J. Kitchen

Born in London, England, in 1913, Father was a metallurgical engineer, specializing in electric steels. He was working at the time at T.B. Balmforth & Company of Luton, Bedfordshire, Boiler Makers and Iron Founders. Grandfather was the General Manager and Father was in charge of the foundry.

At this company, the first war tank was made. Fitted with a Vauxhall engine and transmission, the Mark I was almost impossible to drive. (The top speed was about 5 mph.) Each time one attempted to change gear, the contraption would stop altogether. The original model never went to war but was mounted outside the entrance to Wardown Park in Luton.

Father had demonstrated its capabilities to the War Department by demolishing a cottage donated by Lady Ludlow. Afterwards, the military rejected it as a "useless war machine." One of the observers however was a subaltern named, Winston Churchill. When he became War Lord an order for several improved tanks was placed with Balmforths.

Luton became a city of civil disobedience, caused I expect by unemployment of the demobilized army and the shortage of most things, including food and money. Father was consequently out of work. Thus our home, the splendid "Buena Vista" went on the chopping block. Sister and I were farmed out to an aunt on the Isle of Wight whilst Father and Mother took to the road selling "Lubart Lavatory Disinfectors." Mother did the selling and Father installed the apparatus on the flush pipe of the relevant WC (water closet/bathroom).

Then came a short and disastrous exposure to a Colonel Clark, promoter of the Glen Irving Iron & Steel Company, with plants located in Burton on-Trent and the Port of Richborough. The purpose of the company was to acquire discarded war ships, cut them up and smelt them to pig iron. The ships were located at Richborough, which was the main port of embarkation for the French Front in World War I.

The Colonel did not bother to pay his bills, including the salaries of his officers and men, and absconded with the entire assets of Glen Irving. The firm went bankrupt and the Kitchen family was stranded in the middle of a disused military camp.

I did not object to the situation since this was my element: steam cranes, steam locomotives, power plants and the remains of Glen Irving (smelter, foundry, pattern shop, etc.) and no one around to chase me away. A few small businesses had leased bits of the facility such as sand and gravel from the lake that over looked our back garden. This operation involved the use of the steam locomotives and cranes.

The various operators allowed me all sorts of privileges which would be difficult to obtain today; such as driving, stoking, oiling and repairs. I became quite proficient in my early teens.

Paul D. Andrews

I was born in Lancaster, Pennsylvania, on the 14th of February, 1900. I built my first Amateur Radio Station in 1910. The station did not require a license as DX was about twenty-five miles and it was thought that my signal did not cross any state lines. World War I restrictions closed the station down in 1917.

Aubrey G. Caplan

The inspiration to become an "engineer" came from Alexander Botts and the Earthworm Tractor stories which appeared in *The Saturday Evening Post* magazine. At the age of ten, I read books on civil engineering and learned the names of all the types of bridges.

I also interviewed a prominent local civil engineer and expressed my interest in highways and bridges to him. He was not too encouraging. He told me that with a name like "Caplan," I would find prejudice in this restricted field. He said I would be better off becoming an electrical engineer because firms like "RCA" were unbiased. There would be more chances of a job in the electrical field than in civil engineering. With this in mind, I changed my basic interests to electricity. I learned to repair lamps and electric irons, built a telegraph set and studied basic electricity.

Yardley Beers

I was born in Philadelphia, Pennsylvania, on April 2, 1913, but at the time my parents, Louis Gilbert Beers and Sarah McKim Yardley Beers, lived in Trenton, New Jersey.

I was an only child, and I had virtually no

contact with other children until I entered kindergarten. As a result, I did not know how to behave with other children at first. As a result, I was more interested in things than in people. Moreover, I grew up with an insensitivity to the feelings of others, a deficiency that continued until I reached middle age.

During my early childhood, I was forced to entertain myself much of the time. Soon I became fascinated by trains, both toy electric ones and actual ones. My fascination was enhanced by occasional discussions with my electrical engineer uncle, John Linn McKim Yardley. He, while employed by Westinghouse, had been indirectly involved in the electrification of some railroads.

Actually I found it hard to get Uncle Linn to talk about his work. He had become disillusioned about engineering as a profession. He conceded that it might be acceptable if one were a consultant and could tell others what to do. I replied, "If I really like to work with electricity, why should I tell others what to do instead of doing it myself?" Thus at an early age I learned of the conflicting philosophies of a practicing engineer and of a manager.

William S. Cranmer

In my case, it all started at age six when my father gave me a Mechano set as a birthday present. I learned which way to turn a nut on a clockwise screw-thread and many other worthwhile lessons in the mechanical world. It wasn't long before I was winding my own RF coils on two of my play blocks glued together, and setting up simple radio circuits with the VT-199 and 201A vacuum tubes.

W. Jack Cunningham

From early on, the course of my life was directed toward something technical. My paternal grandparents had moved to a small town in west Texas (Comanche) to escape the devastation brought to their home area of northeastern Alabama by the Civil War. My grandfather set up a general store and, at one time, would go in a covered wagon one hundred miles to the nearest railhead to get items for sale.

Under the circumstances, my grandmother (Harriet Jack Cunningham—she had been Miss Jack and my father and I were named for her) learned to make all sorts of household repairs. Since things were difficult to replace, there was no other way to keep them operating. By the time I was born, she was a widow with little money living in a large, deteriorating house that needed constant attention. When I was a small child she taught me how to use her extensive box of tools, and soon saw to it that I had a toolbox of my own. Before I started going to school, I was making crude wooden models of the World War I airplanes that flew overhead.

My mother's youngest brother (Tom Moore) was a self-taught electrical engineer. On a visit to his home, I was given a magneto from a hand-cranked telephone system. A spin of the crank would produce about one hundred volts ac, but with little current.

Neighbors' chickens were ruining my aunt's flower beds. My uncle decided to solve the problem. He arranged a pan full of moistened bread, supported on insulating blocks above a puddle of water in the flower bed. Long wires connected the magneto to the pan and the puddle. I was to hide behind a bush, wait until a chicken began to peck in the pan, and then crank the magneto. This caused the chicken to leap several feet in the air, screeching wildly, and rush madly off when it hit the ground again. An exciting afternoon was had by everyone.

John J. Dougherty

It all started with a near-fatal case of pneumatic fever on Thanksgiving Day, 1934. Having just turned eleven, my major interests then were football and bicycles, of which I had one and none, respectively.

A fine physician in Paterson, New Jersey, Dr. Peter Roy, knowing that the oncoming teen years must be spent in a semi-invalid routine, searched for some alternatives. "Do you think you might like short-wave radio?" Sounded fascinating. My only "big brother" brother-in-law and New Jersey Bell Telephone sales-engineer, A.P. Wasdyke, found a 6F7-37 ac-dc regenerative receiver kit. With the help of Western Electric engineer-cousin, Gus Pasch, it was soon wired and working.

At first Dad and I listened only to the New York Yankee games and 190 meter police broadcasts. Personally, though, I spent more and more time listening to the "ham-radio" bands and was licensed on April 19, 1938, as W2LHB. (The thrill associated with ever-increasing long distance contacts is still recallable.) Pat Urghart, G4DR, of Bushby, Leistershire and I celebrated fifty years later on September 3, 1989. Despite the wishes of my sainted Mother that she one day would be able to say, as could her Russian-Jewish friend, Mrs. Eigen, "Mein menschen son, the doctor," a career-pattern in radio/electronics engineering was set.

Al Gross

My parents encouraged my work, although they didn't understand it. We were poor, yet my parents found a way to get me parts to work with. The parts may have been from the junk yard, but they kept me busy. At one point, I had our whole house rigged up with radios, so we could communicate from the basement to the other parts of the house. My sister liked that.

At sixteen, I studied hard to earn my Amateur Radio Operator's License, and my parents encouraged me to master the intricacies of Morse code and the other strigent requirements. They weren't technically-minded at all, yet they encouraged me every step of the way.

A. James Ebel

My interest in radio began in the mid-20s when I purchased a Radiola III Regenerative Receiver—DX listening became my hobby. In 1928, I visited radio station WMT in Waterloo, Iowa, my hometown. I was interested in seeing the studio operation; the control board; the big 33-1/3 rpm electrical transcription players; and the transmitter, which was co-located in the control room.

The transmitter used a 250-watt Hartley Oscillator with Heising modulation. It fed a multi-wire flat-top "T" antenna stretching from the top of the Waterloo Morning Tribune building (the location of the station) and an adjacent building. Since there was no frequency control on the Hartley Oscillator, the station's frequency would vary slightly when high winds caused the antenna to swing.

I visited this station regularly because of my intense interest. I would probably have been thrown out as a "pest" (I was a young high school kid) had it not been for the interest that Paul Palmer, an engineer on duty, took. He would even let me come into the control room when things were quiet, and explain how each piece of equipment worked. He also loaned me a copy of the *Radio Manual* written by George Sterling, who, at that time, was head of the FCC Field Bureau.

The *Manual* contained all of the information one needed to pass the examination for a Commercial Radio Operator's License. I studied this *Manual* diligently with a lot of help from Paul Palmer whenever I "got stuck." In April, 1930, I took my Commercial Radio Operator's examination, which at that time was a multiple question essay-type examination calling for explanations of radio technology in general, radio receiver technology, radio transmitter and antenna technology, and the rules and regulations of the Federal Radio Commission. It took me over four hours to complete the examination. I had to draw a diagram of a tuned radio frequency broadcast receiver, and the circuit diagram of a 5000-watt broadcast transmitter. These diagrams were memorized from the *Sterling Radio Manual*.

In May, 1930 (I had just turned seventeen), I received the good news that I passed the Commercial Radio Operator's examination, and received my first Commercial Radio Operator's Certificate. I have maintained my Radio Operator's License ever since.

In June, 1930, I was able to get my first broadcast job at Radio Station KFJB in Marshalltown, Iowa. The vacancy came about because the chief engineer who had designed and built one of the first crystal-controlled broadcast transmitters in the midwest, was killed while doing night maintenance on the transmitter. He had been making adjustments on the antenna and the counterpoise in wet weather. When he came in to make an adjustment on the transmitter, which was in an open rack with no protective circuits, he made contact with 600 volts of battery C-bias, which was fatal. Actually the transmitter was not turned on, but the high voltage bias was in the rack. (A number of engineers were killed in those days at battery-operated stations because batteries are quiet and it is difficult to tell when they are turned on).

The assistant engineer at KFJB, Sid S. Davis (they called him "Steamship" Davis), became Chief Engineer, and I filled the other engineering position. The station had only two operating engineers. I also became the announcer when an announcer didn't show up for the seven a.m. "SIGN ON."

When I returned to high school for my senior year, I obtained a part-time job handling remotes: pipe organ programs from a local theater, and dance remotes from a local ballroom when big bands such as Duke Ellington came to town. I also ran the controls on Fran Allison's first radio appearances on WMT in Waterloo. Fran Allison, later became Aunt Fanny on Don McNeill's Breakfast Club, and Fran in "KUKLA, FRAN and OLLIE" in the early days of television.

I graduated from high school in 1931. In the summer of 1931, I received a job at KGDE in

Fergus Falls, Minnesota. Because this was in the depth of the depression, and advertising revenues were hard to come by, the station had only three employees: the boss, Mr. C. L. Jeren, his wife and yours truly. The station was on the air from seven in the morning until nine at night. I was on duty seven days a week, except that I had an hour off for lunch and an hour off for dinner, during which time Mr. Jeren took over. The operator was also the announcer, who played records, and did commercials by ad-libbing out of the newspaper. In late August, 1932, Mr. Jeren could no longer afford my sixty dollar a month salary, so I decided it was now time to return home to go to go to college. (Actually, this was my father's decision.)

Chapter two

Education

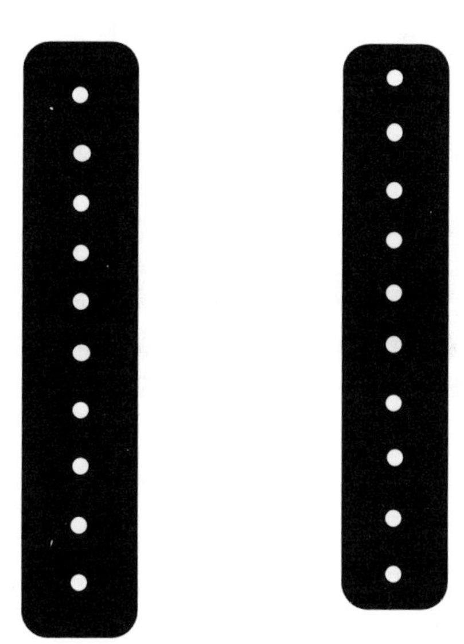

John Duhl

I had no hope for an engineering career as a high school graduate because an invalid father put the family's economic survival in my hands. The war changed things with employment of my sisters freeing me for military enlistment. With the war over and the GI bill available, I tried to pick a specialty which would not be crowded by younger qualifiers. At that time, teaching and engineering seemed to be the choices. By picking physics, either career was possible.

Alexander Lurkis

I decided to become an engineer when I was twelve years old. A pamphlet had been distributed in my class in P.S. (Public School) 52, the Bronx, which described the technical studies in the private non-sectarian Hebrew Technical Institute, located in downtown Manhattan. The possibility of conceiving an engineering project and seeing its completion fascinated me. I took the entrance exam and was admitted upon graduation from elementary school. In the meantime, we had moved from the Bronx to Mt. Vernon so, during my entire high school years, I traveled by the elevated line from White Plains and 241st Street to Eighth Street and the Bowery.

My first technical job upon graduation from the Institute, in 1925, at seventeen, was with the Eighth and Ninth Avenues Railway, first as an inspector, then surveyor, then assistant engineer. Because the company was bankrupt, we had to make do with very little. I had to watch the construction of the Eighth Avenue subway to protect our structures. The project I designed which stands out in my memory was the conversion of a horse-car double-crossing to an electrical crossing, which contained the 600 volt conducting rails, accessible through channel rails in the center between the tracks.

Right after graduation from high school, I took the entrance exam for the Cooper Union Night School of Engineering—a five nights a week/five year course to which I was admitted.

Early in life, I learned that you have to be alert to protect your interests. While with the railway, I had taken a city exam for electrical draftsman and scored high on the list but received no appointment. About a year later, I saw an ad for a new exam for draftsmen within the Board of Education. However, I noticed that the vacancies were being filled by provisionals (political appointments). I immediately filed a protest. I was then hired by the Board of Education Engineering office in Brooklyn. So here I was living in Mt. Vernon, working in Brooklyn and going to school at night in Manhattan.

1930 was an important year in my life for it was the year: I graduated from Cooper Union with a BS in electrical engineering; I passed an exam and was appointed a junior engineer on the Board of Transportation; and, the most important event, I eloped with Carin, an art school graduate of Cooper Union. So, I owe a lot to Cooper Union—for my education and my wife.

Jerome Kurshan

I was always a good student and had a bias toward mathematics and science. Engineering seemed the obvious choice of study for me when I entered college in 1935. However, the prevailing wisdom of the time was that I would face religious discrimination in trying to get a job in engineering. So I took a liberal arts program at Columbia and fulfilled the degree requirements in three and a half years, graduating as class salutatorian with honors in math and physics. This left me with a half year in residence as a graduate assistant taking courses in physics and teaching a lab course to non-matriculated students (in the University Extension program). Professor Robert von Nardroff was in charge of the teaching program and a very fine and capable person.

This was the first of a number of interesting teaching experiences. The common denominator to all of them was what I observed about student motivation. The best learning environment occurs when students are there because they want to be and have a goal in mind. These were adults (all significantly older than I) taking a course after their working hours because they had a personal reason for wanting to learn. Subsequently, when I taught classes of undergraduates, I confirmed this observation. Teaching students who were taking the course primarily because it was a prerequisite in their program was a pretty humdrum experience. Teaching the same subject to a class of repeaters, who had failed it once, was a much more stimulating one. These students, while not as bright, were much more serious about their studies and determined not to fail again.

As an undergraduate, I had taken a course in physical optics which employed a newly published text book. I got in the habit of visiting the instructor, Professor Herman Farwell, after class to discuss the errors that I had found in the text. I suspect that this interaction resulted in a favorable recommendation that helped me get accepted at Cornell as a graduate student and teaching assistant in physics in 1939. The position entitled me to free tuition and an initial stipend of six hundred dollars a year on which I managed to live. Subsequently I was able to stretch my stipend by serving as night watchman, a job which provided me with a free sleeping room in the physics building. This was before the events of World War II and Sputnik enhanced the awareness and image of physics. There were only six of us entering the physics department in that prestigious school that year.

Four years later, I graduated with a Ph.D. degree in physics. By then the war was on and my thesis showed its impact. It was more engineering than physics. It was an investigation of cathode-follower gated amplifiers that had applications to pulse measurement and discrimination in radar. The work was done under the direction of Professor Bruno Rossi and supported by the MIT Radiation Laboratory, which coordinated wartime radar research. Since the work bore a security classification, it could not be published at the time. The required copies were duly printed, bound and sequestered and the university granted me the degree anyway. Several years later, it was declassified and I published the results in *Review of Scientific Instruments* in 1947.

Yardley Beers

I entered Yale University uncertain whether to major in physics or electrical engineering. I made a selection of courses which allowed me to postpone the decision until my junior year. Although I got good grades, I found little interest in the engineering courses I encountered. I, also, was unintentionally discouraged by upper class electrical engineering friends telling me about the long hours they spent on laboratory reports. And I did find physics and mathematics interesting. Therefore I decided in favor of physics.

At Yale, my most important physics teacher was Leigh Page. Pagestated frankly that he did not like quantum mechanics. However, his lectures on the subject were so clear that I remember them better than those of later professors who were leaders in the field. I graduated in 1934, and after a year in the Yale Graduate School, transferred to Princeton where I received MA and Ph.D. degrees in physics in 1937 and 1941, respectively. I specialized in nuclear physics.

Princeton was a wonderful place to study physics with so many distinguished visitors. Among other things, I was in the first audience in the United States to hear about the discovery of nuclear fission. But I was not completely happy there. I had psychological problems. I was fortunate in having Rudolf Ladenburg as a thesis director. He stated, "I never taught you to do research. I solved your psychological problems, and you taught yourself."

Paul Burk

One of my most hilarious experiences was when I held the job of keeping the radios and record players working at the House Plan, a college sponsored social house. They had an early model RCA console with an automatic record changer that must have been designed by Rube Goldberg. After playing a record, the changer would literally toss the record into a side bin of the console. When things were not adjusted just right, the disc might go sailing over the edge of the cabinet and land anywhere.

Also, the console had electromagnetic twelve inch dynamic speakers with more field flux outside the housing than around the voice coil. So when people dropped the steel phonograph needles, they were sure to be found around the field coil. A raspy sound called for cleaning up the discarded needles.

Leslie Balter

Since I started high school at the height of the Great Depression in 1932-33, thoughts of college were minimal. The "big" goal was to get into Townsend Harris High School, and in three years have guaranteed admission to City College, tuition-free. Being one of the lucky ones, I passed the test and was admitted to Townsend Harris Hall.

In those days, Townsend Harris wasn't spoken about much. Today, it is recognized as a prestigious institution with graduates such as Jonas Salk, Herman Wouk (and Leslie Balter).

In high school, I excelled in math and failed foreign languages. I managed to get one hundred on the algebra regents, but had the distinction of failing Spanish (could not remember the vocabulary). This was in spite of having as my Spanish professor, the esteemed Dr. Peter Sammartino, who later founded Fairleigh Dickinson University. His second claim to fame was failing me in high school Spanish.

I secured my automatic admission to City College and, despite the advice of everyone I spoke to, embarked on an engineering career. The engineering profession was a relatively tight, restricted society. There was very little chance a boy from the Bronx could ever break down the religious and the cultural barriers. I had acquired such a love for radio, however, that my father said, "If that's what you really want to do, do it." I DID IT.

As I matured a bit, it became quite obvious that a diploma from City College qualified a person to sell shoes at Macy's (a very good job in those days). I then applied to Columbia Engineering, and was fortunate enough to be accepted. The ultimate thought in my mind was that a Columbia degree might give me a better start than a City College degree. I don't know about the value of the degree, but I did have greater opportunities at Columbia.

In one of my classes, I became friendly with Mal Jennings, who was Chief Operator for Major Edwin Armstrong, the inventor of FM. One of my distinct joys was going up to Alpine, with my buddy Mal, several evenings a week and shutting down W2XMN. Alpine, New Jersey was just a short run over the George Washington Bridge and up Route 9W. W2XMN was the first FM station licensed. At that time, it was conducting purely experimental programs as a method of introducing FM.

The Columbia University Club was active in promoting FM and I worked with a senior student, Bill Hutchins. I recall very specifically putting on a demonstration at Brooklyn Polytech. We coordinated a special demo broadcast through W2XMN to show the quality and interference-free characteristics of frequency modulation.

I was a very essential part of the team because my father was in the garage business, and I had a convertible car at my disposal. The FM receiver was a monster in a five foot rack, and would only fit using a convertible with the top down.

At that point in time, Columbia Engineering School offered two options in electrical engineering. You were either a power man or a radio man. I was in the radio option and we had a total of thirteen students in the class.

Gustav "Gus" Henry Bliesner

My first thoughts for a career were in rural electrification, as electricity should save labor, especially on the farmstead. My maternal grandmother, Anna Rosalie Irmer Stecks, said that I'd be the "doctor" in the family. Ed Clanton's dad, who was an M.D., used to talk about careers in medicine, in addition to doctoring us for childhood diseases. (Back then, the medical profession wasn't practicing vaccination.) Other professions or vocations I was encouraged to consider were the Lutheran ministry, professor and farmer.

I entered Washington State College (WSC) in the fall semester of 1928. I graduated with a BSEE degree June 5, 1933. Fortunately, I made the honor roll five of my nine undergraduate semesters and qualified to enter WSC's graduate school. I made both of the engineering honoraries, Sigma Tau and Tau Beta Pi. I took my MS degree June 11, 1934 and professional Engineering (PE) degree June 5, 1939.

Clive M. Gardam

Although my desire was to major in electronics, at McGill University (Canada) in 1947, the emphasis seemed to be on power. I recall one professor telling us that there were too many students in the electronics option, and he strongly recommended that we choose the power option. "Most of you will end up in the power field anyway as there are too few positions open in electronics," he proclaimed. Apparently, he thought there was a very limited future in electronics! I switched!

Donald Schover

There wasn't money available to send me to a recognized trade school. RCA Institute had a branch at that time in the Merchandise Mart in Chicago where I did janitorial work for part of my tuition. There I received a primitive education in math, physics and circuit theory. Upon graduation, I studied for a first class radiotelephone license and got my first job as an engineer in a small Wisconsin radio station.

Charles R. Smith

At fifteen, I decided engineering was for me after attending an open house at Case Institute of Technology (now Case Western Reserve), and after reading books on the life of Steinmetz and Edison. I had no mentor or family encouragement or financial support, an understandable situation during the depression. However, I still decided, on my own, to get all the subject credits needed to enter Case. I changed my course in high school from general to academic. I got a job on a truck farm at ten cents an hour to earn money for summer school to catch up on foreign language requirements and so forth.

When I graduated from high school I had all the scholastic requirements to enter Case, but no money. Within six months, the U.S. entered WWII, and I worked in a defense plant for twelve hours a day, seven days a week for one year. All my earnings went into War Bonds which guaranteed my tuition for Case if I returned home from military service. I graduated from Case in June, 1950, with a BS in electrical engineering.

Nigel Stace

Examinations are no longer regarded as *sine qua non* in New Zealand. However, fifty years ago, all engineering undergraduates in their final year had to take an examination lasting fifteen days.

This was (hopefully) the last examination. It began after all the other students had left so it was taken in strangely deserted surroundings. The exam was a test of endurance as well as technical knowledge and ability.

Monday to Saturday inclusive, for two and a half weeks, the doors of the Canterbury School of Engineering drawing office and the engineering library were unlocked at nine a.m. and locked at nine p.m. Between those times, the candidates worked unsupervised and were allowed to search for and share information from any source that would enable them to design a specified product, structure, or machine. Then they had to make appropriate drawings of it.

Preparation for this ordeal formed a complete subject in the final year and, in the case of electrical students, it was always for the design and drawing of an industrial motor or generator—since these were usually requested. But the examiners, who resided in England and not necessarily *au fait* with what was used or taught in New Zealand, sometimes changed the request. This happened in 1936—when five horrified candidates found themselves asked to design a motor for an electric railway locomotive, rare things in New Zealand but common in England.

Such motors involved some unknown design parameters, since the railway track, traction motors had to be so restricted in physical size that few of the normal design criteria applied. In short, they had to be designed way beyond the textbook's electromagnetic limits.

A frantic search of the engineering school library and all local book shops found no books or journals on appropriate design. The search also used up some precious time.

It was then resolved that, working as a group and sharing all information, the candidates would extrapolate from existing design information, make hopefully informed guesses where necessary and come up with one composite design. They would rely upon their differing drawing abilities to differentiate their exam entries.

Fortunately, tradition had it that no one had ever failed the marathon fifteen day test. After one hundred and eighty hours of frantic work and many sleepless nights, the designs' drawings were completed and handed in. Then there was a lapse of over two months while the entries went surface mail to England for marking and the results cabled back.

The composite design had obviously been mediocre; but the quality of the drawings mostly adequate. All marks were disturbingly low, but only the most untidy draftsman failed and had to repeat the entire exercise the following year.

Perhaps more fortunately, electric locomotives for New Zealand continued to be designed and manufactured overseas.

Philip Sproul

I attended Iowa State University (ISU) for two reasons: I could afford it (almost), and it was a well recognized engineering school. I was one of about three in our electrical engineering class of thirty-eight students interested in communications. As a result, I got a lot of attention from the only professor in this field.

I worked in the summer between my junior and senior year (1936) as a construction worker for Northwestern Bell in Des Moines. Jobs were few and far between in this depression period. In early 1937, I was interviewed by a Bell System team, recruiting for the first time in many years. Bell Laboratories had never recruited ISU before. I had never heard of Bell Laboratories. However, the head of the electrical engineering department told me about it and said that was where I should go and that I would get an offer. I did, accepted it and retired in 1980 after forty-three years.

Gregory Timoshenko

I do not recall the existence of "elementary schools." In the schools I knew, one entered when one could read, at perhaps the third grade level, knew some of the multiplication table, and so forth. The schools were designated either as "normal" (eight year) or as "gymnasium" (twelve year).

The normal schools had eight years of mathematics, while the gymnasiums required more in foreign languages, including the study of Greek and Latin. In fact, all subjects in all the Russian schools were required, so a number of students never graduated. This did not result in rejection by society, however. These "dropouts" still could have gainful employment—one was paid for what one could do rather than on the basis of a certificate of education.

Grandfather, who was originally a son of a serf and grew up as an orphan, had a good size abacus, about ten inches by twelve inches with large yellow beads, while numbers five and six were black beads. He could operate it at high speed, multiplying perhaps 1.24 by 85 in a few seconds. I watched him do it and he let me push some beads counting to five or six.

By 1920, the Soviet government eliminated the last two years of high school study, so I graduated at the end of the sixth year. The government wanted more young people in military service! Other changes had taken place: a "student council" was formed, of which I was elected chairman. We made the curriculum, set the teachers' pay, and set the schedule for whose turn it was to bring wood for the small potbelly stoves we installed in each classroom. (The central heating system had broken down.) To get wood in a large city, one had to break somebody's fence! As a result of developing such habits, during Christmas of 1919, I dragged home a beautiful Christmas tree which I had cut down in a public park! Mother was horrified, but one could not put it back.

Upon our escape from Russia in May of 1920, we ended up in Zagreb, Yugoslavia, the part that had belonged to Austria before the 1918 Treaty of Versailles. The school gave me a year to read and write in Croatian, learn their history and catch up on six preceding years of drafting and geometric problem solving ("descriptive geometry" in several modes of projection!). I frequently had to study seven days a week, but finally I "made it." After the two weeks given to students to prepare for final examinations, I completed a week of daily four-hour written examinations and an oral examination before a board of state examiners.

The diploma was called "the Certificate of Maturity." Courses included two years of calculus, two years of physics, and two years of chemistry (inorganic & organic, using two four hundred page university level volumes). Five sections of forty students each started in my class; after four years only twenty were selected for the fifth year, and only sixteen earned their diplomas!

Students who made it through the first four years (around eighty) went on as apprentices into various business. Those who earned the eight-year diploma were admitted without entrance examinations to the most prestigious engineering schools of Europe. I went to the "Technical University Berlin" in Germany.

H. R. Weiss

I arrived in the United States in 1940, a refugee from Nazi Germany. A year later, my parents and I moved to Massachusetts; I was a little disappointed because, living in New York, I had hoped to study engineering at the night school of NYU or the City College. At that time, there was no way to receive a Bachelor of Science in engineering going to night school in Massachusetts. (One could get a master's if one had a bachelor's, but that didn't apply to me.)

A refugee committee informed me that someone was willing to sponsor my education. However, when I said I wanted to be an electrical engineer, they replied that they had rather thought in terms of an automobile mechanic or body repairman. Then it was suggested that, if I really wanted to study engineering, I could do so at the Lowell Institute School, a two year technical evening school at MIT which was free. This decision to shortcut my education must have eventually cost me many years of intensive study, but I was too inexperienced to see the final outcome.

Warren L. Braun

Although all of my instructors at Valparaiso were excellent, I remember most favorably Dr. Joseph B. Hershman, Dean of the School, who also taught propagation theory and acoustics. His first day in the classroom was most memorable. He was a big man, overly corpulent, wearing a dark blue pinstripe suit. A Phi Beta Kapa key dangled very visibly from a watch chain draped across a tight fitting vest.

On the first day of instruction, he charged into the classroom announcing in a stentorial voice, "I am Dr. Joseph B. Hershman, B.S., M.S., and Ph.D. Before you get carried away with the initials, I remind you that B.S. is best left to the cows from which it came, M.S. is more of the same, und Ph.D. means piled higher and deeper. Now let us go to vurk." With that, he literally tore off his coat, rolled up his sleeves, laid the watch and the Phi Beta Kappa key on the desk, and lectured at a furious pace usually breaking chalk until every board was full.

Joseph F. Furlong

Eventually, in the fall of 1938, I entered Union College as a freshman in the electrical engineering curriculum. Union is primarily a liberal arts college and we were a small group of about twenty-five aspiring "electricals." The electives in the electrical engineering curriculum at that time were few; in fact the elective was: "Take it or leave it."

The head of the electrical engineering department was Dr. Ernst Julius Berg. He had a habit, which caused him a great deal of embarrassment but which we found amusing, of talking to us with his ever present pipe between his teeth and (unconsciously) blowing through it. This created a fountain of ashes that landed on whomever he was talking to. Dr. Berg had been a close friend of Steinmetz, after they met working at General Electric (GE), and he also had known Oliver Heaviside whom he had visited in England.

As a result of that visit, he wrote a textbook on Heaviside's operational calculus and the application of its "unit function in electrical engineering." Because he felt we needed it in addition to the mathematics we were being taught, we received liberal doses of Heaviside in Dr. Berg's electrical engineering classes.

Unhappily, the summer of 1941, between my junior and senior years, Dr. Berg died unexpectedly. The college was able to engage two young GE engineers to teach the power and communication courses, those being the two sides of electrical engineering at that time. The task of teaching us "power" was taken on by Charlie Kilbourne and the "communication" teaching fell to Simon Ramo. It was a tough year for us, with six eight o'clock classes in electrical engineering every week, but I am sure it was tougher on our teachers, Kilbourne and Ramo. As the years have passed, it has become increasingly apparent how much they did for us that year.

R. H. Eberstadt

Again, remember this is the time of the world wide depression, and where we lived, in Mexico City, there was not much of a choice of colleges. Therefore, where I could get the best support towards "scholarship" is where I turned to, and it was an excellent choice at the time—the National Autonomous University of Mexico. While going to college, I became interested in another area of engineering: aereonautical engineering; but the career did not exist at this university at the time. To make the story short, when I graduated, I helped establish this career at the college, unfortunately it later languished and finally disappeared.

J. Coleman White

When I was studying at Cornell University, many years ago, I had a favorite professor—Everett M. Strong—who specialized in lighting. One afternoon, as I headed home from an electrical engineering laboratory, I ran into him in the hallway. After a short chat, he asked me if I had a few minutes. He had something he wanted to show me.

I had the time, so he led me to his laboratory where he had the following relatively simple set-up. There was a conventional movie screen. About eight feet away from it, there was a battery of light projectors, consisting of a central projector surrounded by six others. In between was a black circular piece, probably cardboard, supported on a thin metal rod. When any projector was on, the screen would be cast in light, except for a circular shadow caused by the intervening piece.

He began by turning on the central projector, which produced essentially white light. The screen was then quite white, and the spot quite black. Then, he turned on a second projector, which had a very faint gelatin of red color over the lens. The amount of light coming from this second projector was adjusted to be quite low. What happened when this second projector was turned on was that the screen didn't appear to change at all. It still looked completely white. However, the change in the spot was truly remarkable! It had turned a bright green color. When I gasped at this, I was handed a cardboard tube, and I was told to look at the spot through it. I did so, and found that it was quite black! The apparent green color of the spot was only in my mind. It recognized the slight reddish tint of the background and inserted the complementary color on the spot.

Each projector had a gelatin of one of the primary or secondary colors, and the process was repeated for each. In each case, the spot appeared to be brilliantly colored with the complementary color being projected.

Then, as sort of a finale, Professor Strong turned on all seven lights. What resulted was, again, a seemingly perfectly white screen, but this time a spot that looked, more than anything else, like the rose window in a cathedral. It was covered with arc shaped segments, each of a different brilliant color. The delineation between the segments was quite sharp. However, a look through the cardboard tube confirmed that, in reality, the spot on the screen was totally black.

This, of course, led to an informal lecture on the meaning of what I had seen. Several points were made:

•When we look at a tree, we see the green leaves, but we also, through our own minds, put red (slightly) in all the shadows cast by each leaf. The child not knowing this uses only green, so the tree looks unnatural The professional artist knows that it is necessary to put in the red, although he/she may not know why.

•This is why the worst color of golf tee to buy is dark red. You will stand a much better chance of losing it than you would a yellow or white one, for example.

•Another common case is shadows on snow. This is easy to test out. The sunlight is slightly amber in color. This causes us to color the shadows on the snow a definite blue even though a look through a tube, to block out the sunlight from view, would confirm that the shadows are really dark gray or black in color, and not blue at all.

The moral of all this: When we look at something, how much of the scene is physically there, and how much have we modified?

Thomas A. Nelson

By the time I graduated from high school, I felt that finances were such that I could not make it through four years of college. The solution was to enter the semi-professional electrical engineering program at Los Angeles City College (LACC). The college had marvelous machinery labs, both electrical and mechanical. While there, I served as president of the Engineering Honor Society (EHS).

World War II was upon us several months before I received my Associate in Arts degree. I signed up for the aviation cadet program in the Army Air Force, requesting communications training. This seemed to be the closest option to my engineering training.

A few months later, I was notified that communications classes were full, and I was assigned to the armament option. I was told to report to aviation cadet basic training at Valley Forge Military Academy. It was winter then, December, 1942. As a Californian, I came to appreciate what George Washington must have experienced at Valley Forge.

The armament studies commenced at the aviation cadet advanced school at Yale University. Eventually, I found that armament wasn't too bad a choice, because it involved electrical circuit applications, such as in the bomb release system and gun turrets on the aircraft. After graduation from Yale as a second lieutenant, I served in B-24 Liberator bomber groups in the European Theater, reaching the rank of captain before VE Day.

As soon as the GI Bill was announced, I decided to return to college. I went back to LACC and finished the lower division engineering curriculum requirements and reestablished the dormant EHS, becoming its president again. I transferred to the University of Southern California (USC) for upper division work.

At that time, most of us in university engineering classes were above the usual student age, some with families, utilizing the GI Bill, and not afraid to express ourselves. I recall one particular incident in a class at USC that illustrates this point. It was the day before Thanksgiving recess, and the instructor placed on the chalk board a particularly heavy homework assignment.

One of the students in the back of the classroom, in a disturbed, loud voice, said to the instructor, "What do you expect us to do, work on our assignment with one hand and eat turkey with the other?" Still, the assignment was not withdrawn. Well, most of us survived the rigors of academia, although there were a significant number of engineering students who became business administration majors.

Francis J. Heyden

In 1938 I applied for admission to Harvard to work for a doctorate in astronomy. I was very green and, with four full courses in atomic physics, statistical astronomy, astrophysics and research, I worked longer and harder than ever.

The first qualifying examinations came after six months. I reached bottom with a fifteen percent in one. My advisor thought it was wonderful joke. In the second year I obtained a fifty-eight out of a passing sixty percent. This was a warning. Then in my third year I was ready to give up. For days after the tests, I thought of the answers I should have written. Then one afternoon Fred Whipple sent for me. He shook my hand and congratulated me on the wonderful results. I still save those papers. One has Harlow Shapley's remark, "This is the best examination paper I have ever seen." Humility goes a long way towards getting results.

John Alrich

The decision I had to make about my lifetime career was whether it would be in engineering, mathematics or physics. Since in those days physics and math majors were required to be conversant with a foreign language (usually German) and I disliked learning languages, engineering became my choice almost by default. I use the term "almost" because the other positive factors in selecting engineering were its utility, pragmatism and immediacy of application, often showing positive results in a relatively short time span. (Perhaps something much less true today than it was forty years ago.)

I graduated with a BS in electrical engineering from University of California-Berkeley in June, 1948. Jobs were easy to find and most of my friends and I had several offers before we graduated. I look back at the last two years I spent at Berkeley as being among the happiest, most care-free (the GI Bill paid nearly all my expenses), stimulating periods of my life.

The instructor I remember the most vividly was John R. Whinnery, who was teaching electromagnetic theory while he completed his Ph.D. He didn't know it at the time, but if there was anything that discouraged many of us from going on to graduate work, it was John! He was about our age but had already published in the *IRE Proceedings* (as it was called) a number of times and co-authored a textbook with Simon Ramo. Quite a few of us thought that if one had to be as bright and disciplined as Whinnery to go beyond a BS, there was little hope for us mortals.

Also I was twenty-five years old by this time and anxious to get into industry, marry raise a family; i.e. plunge into a rather conventional life-style that had been denied most young men like myself, recently returned from the service. Although over the years I did do graduate work from time-to-time, I never did get back for an advanced degree. This may have been a mistake, but in those days advanced degrees were less necessary than they are today.

Thomas M. Austin

I was in high school when I first heard about electrified railroads and dreamed of being the engineer to electrify the big western railroads; however, I never did that.

I grew up on a farm. My father encouraged me to enter the electrical field after we got electric service on the farm about 1920. We first considered a trade school, but settled on an engineering education. We lived in southern Colorado, and decided on the University of Colorado at Boulder because it had the best electrical engineering program in Colorado and was conveniently located. In 1932 another senior, Fred W. Cooper, and I wrote a paper on non-linear magnetic circuits. This paper won the 1932 AIEE National Prize for a student paper, and was later published in the *AIEE Journal*.

Aubrey G. Caplan

In the fall of 1941, after finishing high school, I began college at Carnegie Institute of Technology (now Carnegie Mellon University) but soon ran out of money. It became necessary to earn enough money to buy books, to eat lunches, and date girls. I took a job at a supermarket working from four p.m. to ten p.m. each night and all day Saturday for twenty-five cents an hour plus free cigarettes. This lasted until January since I was doing most of my sleeping in class. I also repaired appliances. I remember charging two ladies five dollars for plugging in their refrigerator which had become disconnected. This was not too exorbitant a charge, since it took two hours travel time by streetcar to reach the job site.

Sidney Bertram

I was born Sydney Abramovitch in Winnipeg, Canada; the name was changed when the family moved to California in 1923.

I took a vocational electrical course in high school graduating in February, 1930. The only job I could find was on a radio assembly line where I worked on as many as two hundred radios a day at three cents a radio.

By fall, tired of this work, I enrolled in the semi-professional electrical engineering program at the Los Angeles City College (LACC). I graduated in 1932, with an Associate of Arts degree, an excellent foundation for my future work, but with no job and decided to continue my education. I entered Cal Tech as a junior (by examinations in math and physics) but left after about two weeks because of money problems and difficulty in registering for freshman chemistry. (I hadn't planned to go on to college from LACC.) I returned to the junior college in 1933 to make up the chemistry, but a course in radio altered my plans.

In the fall of 1933, I entered the Radio Institute of California in Los Angeles. I was soon offered a part time job as a laboratory instructor and later, when the theory instructor left, I was invited to teach beginning radio theory which gradually became more advanced. I stayed there until the fall of 1936 when I left to re-enter Cal

Tech where, in 1938, I obtained my BS (with Honors) in electrical engineering.

After a year with a small geophysics company, I enrolled as a graduate student at the Ohio State University (OSU) where I received my MS degree in 1941. My thesis on the "Calculation of Axially Symmetric Fields" was published in two papers that have been referenced in the literature on electron optics. (My advisor offered to let me use it for my Ph.D. dissertation, but I was then too far from being able to satisfy the language requirement.)

Following the war, I worked for one year at Boeing primarily on the guidance elements for a "ground to air pilotless aircraft." I left there in 1946 when I was offered an assistant professorship in electrical engineering to teach at the Wright Field Graduate Center of OSU with the opportunity to finish my Ph.D. I completed it in physics in 1951.

William S. Cranmer

Formal education was uneventful except for the unusual circumstance that I received a BS from Rutgers University in accounting. I hesitate to mention this fact as the IEEE may question my long standing status as "Senior member." After four years of Gun-Laying-Radar work during World War II in the European Theater, I had no desire to enter a career of accounting and in 1945 joined the RCA Electron Tube Division in Harrison, New Jersey.

Following six years of night school at Stevens Institute of Technology, I finally earned a MS degree in engineering management in 1951. It is interesting that at one point a school official said, "We really don't like to award a master's degree to a business student, but you passed all the courses, so I guess we have to go ahead....."

Louisa S. Cook

After one semester of typing in high school, the instructor recommended I switch to math. My typing was so poor that I would starve as a secretary.

I became an engineer because my first job, after high school graduation in 1942, gave me experience in engineering which I thoroughly enjoyed. Six months later, I won a small college scholarship for 4-H work I had done for my folks during the depression and on through high school.

My co-workers at the U.S. Bureau of Reclamation encouraged me to become an engineer since they had a woman in their engineering classes who had graduated and entered the field. I had the good fortune to meet that pioneer, Jane Rider, when I finished college. The Yuma Union High School mathematics instructor who recommended me for this first job, when my male classmates and men from U.S.B.R. were being drafted for World War II, did the most for getting me started in engineering.

I was the first member of our family to attend college. I chose the University of Arizona in Tucson because it had the only engineering college in Arizona at the time (1943-1947). Also it had a cooperative dormitory for which my 4-H work qualified me as a resident. There work in the dormitory greatly reduced living expenses.

W. Jack Cunningham

After graduation from Waco High School in 1933, in the depth of the depression, I attended

Baylor University for two years because my mother was able to promote a tuition scholarship. Also, I could live at home with minimum expense. During this time, I worked afternoons in a shop repairing home radios, receiving my lunches and three dollars a week.

In 1935, I transferred to the University of Texas (Austin), and majored in physics. I lived in a dormitory with its own dining hall where the monthly fee for room and board was thirty dollars, rising to thirty-three dollars the last year I was there.

I worked as an assistant in the physics laboratories. We were paid our eighteen dollars a month with checks that were not valid when we received them. Only after the state legislature made necessary appropriations, following several months' delay, were the checks cashable. Meanwhile, there were small establishments which made a business of cashing the checks at a discount. The last year I was in Austin, I was paid four hundred and fifty dollars by the University, now in cashable checks. I was able to meet all my expenses for the year and have a little left over. I stayed three years, receiving both the bachelor's and master's degree in physics.

I was impressed by a young physics professor, Paul Boner. Boner had done his graduate study at Harvard University. The second year I was at the University of Texas, he managed to have two of his students accepted by Harvard for graduate study. The next year he asked me if I would like to go to Harvard. Naturally, I said yes I wanted to go. Wi and recommendations from him and from his colleague, Arnold Romberg, helped me get accepted there.

Edgar C. Gentle, Jr.

Did you have a mentor?

Yes, I expect I did like most people starting out. There was an older engineer with whom I worked when I was going to college. And, incidentally, I was in Auburn's first cooperative engineering class and spent my summer and winter assignments working with Southern Bell Telephone and Telegraph Company. I worked during the winter quarters, went to school in the spring quarters, and worked during the Summer Quarters starting my freshman year in 1938.

My choice of college was based on the reputation of Auburn University. In those days, it was the Alabama Polytechnic Institute and had an excellent reputation in the U.S., particularly in the South, and in the Bell System (in fact a degree from Auburn was said to be almost as good as a college education.")

AT&T and the Bell Operating Companies have had many presidents who were Auburn University Engineering graduates. South Central Bell, Bell South, Bell Laboratories and AT&T Long Lines also have had many officers and other high executives, as well as presidents, who are or were Auburn University/Alabama Polytechnic Institute graduates.

Yes, I've continued to do much self-study. Also, AT&T South Central Bell and Southern Bell saw to it that my broad education was a continuing process. I've gone to such schools as Williams College and Columbia University and have studied such subjects as Economic and Constitutional History of United States, Foreign Relations, Diplomatic History, History of United States Supreme Court, Finance, Pricing, Marketing and Sales Management. It does take continuing education. I've also been involved in some of the very advanced technical programs particularly at the middle and upper management level in the Bell System. There were many opportunities presented not just to me but to many others.

William A. Edson

The University of Kansas (at Lawrence) was close to my home (Olathe) and had a good school for electrical engineering. My father, brother, and several uncles had gone there, so it was the obvious choice. My years at KU were the bottom years of the Great Depression; so when I graduated in 1934 there just weren't any jobs. Out of a very talented class, only two electrical engineers got job offers. So, with the help of my family and two part-time jobs, I stayed at KU another year to get a masters degree.

Near the end of that year, Dean George Shaad called me into his office and offered me a Gordon Mackay Scholarship at Harvard! For that I am deeply grateful to him and to Conyers Herring, who had received the same honor the preceding year, and who was already showing his outstanding talents at Harvard.

Glydus Gregory

As a child, or young man, I dreamed of being a doctor, a writer, an actor, an auto mechanic or a soldier—General, of course. But I spent too many years in the fourth grade. I finally did graduate from the eighth grade at the age of seventeen after attending a total of thirty-six months in five different schools over a ten year period.

I studied auto mechanics but could never get a job. My studies were frequently free, or at a low cost, as I seldom had any money. In 1938, I won a scholarship to the University of Wisconsin and attended their school of economics that summer. I also worked on several WPA projects during the thirties and early forties, including my machine-shop studies. I studied machine shop and did get a job but was told by my bosses that I would never make a good machinist.

All through the thirties, I had been attending night school and had attended a few classes on radio and had joined a ham club and built a number of radios. About 1940, I had been given a correspondence course from the American Radio TV school in Chicago, run by Colonel Senabra. He and a night school teacher named "Rainence" had impressed me by their lectures and writings.

On the strength of this, I applied to the U.S. Signal Corps for their Electronics Training Group. They sent me to Chicago and Philadelphia for training and then sent me to California, where I worked on radar most of the time until after the war.

I continued to study and, in 1945, I passed the FCC exam and obtained my first class radio-telephone operator's license. I later added a radar endorsement. Then in 1946, I went to work for the California Division of Highways, later known as Cal-Trans, where I worked as an technician and engineer for twenty years. I retired in 1974 as a Electronics Telecommunications Engineer.

While working for the state, I continued my night school studies. In 1955, I received my high school diploma and in 1966 I graduated from Sacramento City College with an AA degree, with a major in science.

William Hughes

Mr. Hughes has been known to me for more than twenty years. He was a student in the Technical High School for five years. During that time he showed great aptitude for engineering work and passed the University Entrance Examination. He was a senior non-commissioned officer in our cadet companies totaling six hundred boys and also passed the examination for commissioned rank. He was a prefect for three years and head prefect of the school for two years.

For at least a year before leaving the school, he was a student teacher in our Mechanical Engineering Department and in this work he showed the same sense of initiative, dependability and responsibility as he did in all his other school activities.

Mr. Hughes left our school to enter the employ of the Municipal Electricity Department, and whilst there he attended the Engineering Evening School at Canterbury College.

About 1929 he was appointed instructor in electrical engineering in the Evening School and he has filled that position most satisfactorily ever since. His classes have been well attended and, despite the war, the number of his students, mostly electrical engineering apprentices, has increased year by year. He has shown good results from his classes in examinations for electrical wiring and in City and Guilds examinations.

During the past ten years he has been a full-time member of the Technical High School staff, taking electrical engineering, electricity & magnetism and mechanical drawing. He has put all his ability and energy into his work and has given the boys a sound and thorough training.

He is in charge of the Electrical Engineering Department, with oversight also over the instruction given in the radio service classes in the Evening School. He is responsible for preparing the syllabuses of instruction in his department and for the care and purchase of its equipment.

His discipline is good and he is on happy terms with pupils and colleagues alike. He conducted a radio club out of school hours for several years prior to the war and many of its members qualified for transmitting licenses. He has also been in charge of the signaling section of the school cadet battalion for many years.

Mr. Hughes' connection with the College in all its phases has been a highly creditable one, and it has been one from which the College has profited. He is a man of sterling character and an old student of whom the College has reason to be proud.

D. E. Hansen
Principal, Christchurch Technical College
August 8, 1944

C. R. Schmidt

An employer will almost never transfer a senior man to a new product line involving new technology. In this situation, the employer invariably chooses a junior or novice engineer. His reasoning is simple, both the senior and junior man will take about the same time to do the job and the junior man gets much less pay. Hence, there is an obvious cost saving. No age discrimination here, just basic economics.

Engineers of all ages should look toward retraining themselves as new technologies emerge. Companies—particularly the smaller ones which employ the bulk of engineers—are more fragile than they appear. Sometimes they become victims of unimaginative marketing; sometimes they are bought and restructured; sometimes the original owners die. In any event, companies do go out of business. Now a pension bound engineer who has not kept abreast of some new technology can be in big trouble. This disappearance of jobs probably accounts for the largest disappearance of engineers from active practice.

An engineer's retraining begins with oneself. The emerging technology which appeals to him or her should be the first consideration. Without a special interest, nothing will happen.

Out of my interest in writing programs for the Intel 80C48, I acquired a Commodore VIC-20 at a price of sixty-nine dollars, a twelve inch black and white television at sixty-five dollars, a Promqueen at two hundred dollars, and an OKI development board for simulating an 80Ç48 at about fifty dollars. I set myself the goal of writing a serial keyboard routine that would minimize the number of In-Out ports required of the 80C48.

It took me two months to get the hardware together and another two months to assemble it and write the program. The day finally arrived when I could press any of sixteen keys and have its symbol come up on a liquid crystal display, a fairly conventional result. However, I had accomplished this with only three In-Out ports of the 80C48 instead of the conventional eight ports.

Well, even though I had retrained myself in this field, my company brought in a junior man when they decided to go into microprocessor designs. This young man, twenty-seven years old, got his experience at Eastman Kodak in Rochester. In a few more years, he will have to think about retraining himself.

As far as I am concerned, I know I am competent in the microprocessor field. When the time comes I will prove it. After all, at the time I was only sixty-six years old.

Charles T. Morrow

During my senior year, my faculty advisor in physics was Professor Frederick (Ted) Vinton Hunt, a blossoming pioneer in electronics and acoustics. Partly through his sponsorship, I found that I could obtain a scholarship in the graduate school of engineering.

Compared to the physics department, to which I had been attracted, this made the difference only of one required course in electrical machinery rather than an extra course in quantum mechanics. In most of the courses I took, physics students and engineering students attended together with no obvious distinction. (This would give me a special vantage point many years later, when I attended meetings of the American Society for Engineering Education.)

Accordingly, I set out for degrees in electronics and acoustics. In the course of time, I was to do a significant amount of research in acoustics. In my work in aerospace, acoustics turned out to be an excellent bridge field between electronics and mechanical engineering. It included no information on fatigue or other mechanical failure mechanisms, but it permitted transferring powerful theorems from electronics to mechanical engineering—the applicable differential equations are very similar. Furthermore, most of the instrumentation for mechanical engineering problems rapidly became electronic.

In this way, I came to study under a number of distinguished pioneers of physics and engineering. The teaching at the Harvard Graduate School of Engineering was not obviously better than possibly at a more average institution, but this association provided a more realistic idea of what true pioneers were like. There was no valid stereotype. Many of the professors made solid achievements in spite of definite shortcomings in their approaches, and some because of such shortcomings. In any event, as the years went by, it was easier for me to make my own contributions unburdened by over idealistic impressions of the type of personal character required.

Ted Hunt was proficient in experiment, theory and invention. He was frequently bothered by attacks of asthma, which he told me was the result of stresses he experienced while a student in graduate school. He was a colorful character, driven hard by competitiveness and ambition, and inclined to insert himself in his students' lives more than he probably should have. (Something I did not entirely understand before the memorial session held for him in Cambridge by his students.)

We all had similar problems with him, but I think none of us would want to have missed the association. Frustrated by the youth culture that developed among the students in the sixties, Ted retired a bit short of his entire tenure. Shortly afterward, while attending a meeting of the Acoustical Society of America, he suffered a heart attack and died.

The suggestions that Ted gave me for my thesis research amounted to an assignment that was probably impossible—first, the solution of the sound wave equation in an auditorium of arbitrary shape. Along the way, he placed great emphasis on the design of a resistance-inductance oscillator. This would have the advantage over the popular Hewlett Packard R-C oscillator because thefrequency would be proportional to resistance. In addition, he wanted it to be frequency modulated for better use in measurement of reverberation time.

This capability, easy to achieve in a beat-frequency oscillator, proved difficult to incorporate into a circuit operating entirely at audio frequencies. We soon encountered a conflict of approach. When Ted faced a roadblock, he would start a wild series of experiments in all the directions he could think of. This worked for him. My instinct, at the opposite extreme, was to stop and meditate.

Ted read more interpretation into my difficulties than was valid. My most basic difficulty was quite simple. I had missed the social life of college. In graduate school, I was more interested in catching up—associating with students in other fields rather than getting to know intimately the students in my own field and what they were doing. In the long run, this was good. (Graduate school can be a narrowing experience.) But I began my thesis research with little knowledge of how my fellow acoustics students were approaching their own and, for the time being, I was isolated in my own field. I did not know how to slice out a significant but soluble problem by myself. This was resolved by World War II.

Before long, Pearl Harbor was attacked. I

will never forget the front pages of the Boston newspapers a day later. They were blank except for two headlines. The first, in large bold type, said "HITLER DECLARES WAR ON THE UNITED STATES." The second, below in small type, said, "Me too says Mussolini." No one studied in the dormitory that night. We sat and talked. Although I did not know what would become of me, I felt relief from the helplessness of watching Hitler's ruthless advance through Europe.

About a year before Pearl Harbor, Harvard began to tailor its functions to probable wartime conditions. Ted hired me as the first full-time employee of the Harvard Underwater Sound Laboratory, where I spent several months designing underwater microphones.

Then I was transferred to MIT for several months to work on acoustic mines under Cyril Harris and Professor Philip M. Morse. When it turned out that what we were trying to do had already been accomplished at another laboratory, I returned to Harvard to teach a pre-radar course and an electronics course for Navy communications officers under Professor Emory Leon Chaffee.

When the Navy took over the latter course, I joined the Harvard Electro-acoustic Laboratory directed by Dr. Leo Beranek. I stayed there for the remainder of the war. However, as an emeritus of the pre-radar program, I continued to enjoy social events such as faculty-student parties and the nights we bought out the floor of the Pops Concert at Symphony Hall.

I finished up a small project on communication between divers, started by other members of the Electro-Acoustic Laboratory. The bulk of my effort, though, was on speech communication through gas masks. I was told initially that no one had made any progress on this before. My attitude was, What do I have to lose?

It turned out that the previous investigators had treated the problem as an extension of microphone intelligibility—purely a signal transmission problem. In other words, they neglected the interaction of the acoustic impedance of the mask with the impedance of the vocal tract. By placing primary emphasis on the reaction of the mask on the voice, I achieved a remarkable degree of success.

When the war ended, I proposed to supplement my work by a theoretical and experimental investigation of vowel distortions produced by cavities attached to the mouth. I would submit the total as a doctorate thesis. Ted took the position that graduate students who had been employed at the war laboratories would be better off going directly to industry without the doctorate. When the first case came up for review by the faculty, he was overruled. That set a precedent. In this way, I came to get my doctorate degree under the only professor who opposed my research. However, after my final oral examination, he was very complimentary.

Until the last few weeks before the due date for the thesis, my experimental work was unrewarding and frustrating. There were as yet no magnetic tape recorders. I had to use a Miller recorder, which in principle combined the best features of phonograph and film, it recorded mechanically but reproduced optically. However, a trained technician was not supplied with the machine to keep it working. I finally completed my voice recordings and made my harmonic analyses with a wave analyzer. My experimental results at least followed the trend of my theory.

Kenneth G. McKay

English speaking Montrealers rarely shopped around for a suitable university; they simply went to McGill. Like Everest, it was there. In 1934, I enrolled as an engineering student. However, I soon found out that I would be forced to spend many hours at a drafting board performing "mechanical engineering," an unfortunate ambiguity. Reasoning that when I became a successful engineer I could always hire draftsmen, I shifted to the Honors Maths and Physics course.

The Honors Math and Physics course was unusual. After freshman year, the required courses consisted solely of various aspects of mathematics and physics. A correlation was the class size: at first we were three, later two. Being head of the class was strictly a binary function. It was not too difficult to graduate summa cum laude with a B.Sc. degree and the Gold Medal for Natural Philosophy.

After much negotiation, the Student Athletic Council was formed in 1938 to which I was elected president. Great were my expectations. Plans were designed to extend the scope and spirit of athletics at McGill which I presented to the faculty members constituting the Athletic Council.

Unfortunately, the implementation of our plans would cost money; my charter did not include any access to McGill's financial resources and the bursar was entirely deaf to my pleas. Our only leverage lay in moral persuasion with which, hopefully, subsequent S.A.C. presidents had more success than I. Lesson: purse-string control is vital to the success of a project. Probably, that contributed more to my education than any other single event at McGill.

I had been granted the Moyse Fellowship—the prize award which was usually exercised at Cambridge University, England, or under unusual circumstances, at the University of California, Berkeley. I informed the head of the McGill physics department that, on the basis of some library publications, I had decided to go to MIT. He was appalled, "How could you squander the Moyse on a trade school?" Clearly, he had not caught up with the MIT "Compton Effect." (Don't misunderstand me; I received an excellent education at McGill. It was not the only university to contain some unusual characters with parochial blinders.)

I arrived at MIT just as World War II broke out in Europe. My credits from McGill were accepted completely so I only had to take three brief courses. I could devote the rest of my time studying for the orals and carrying out my doctoral thesis. My thesis professor, whom I had selected from articles in the *Journal of Applied Physics*, was an excellent experimentalist. He was meticulous in his own measurement techniques and attempted to convey this philosophy to his students.

However, with some it didn't take: when I was ushered into the cell-like room that was to be my lab, the floor was strewn with broken glass and a great deal of liquid mercury. The previous student, nearing the end of his thesis, suspected a leak in the elaborate glass vacuum system he had constructed. He proceeded to search for it with a high voltage spark, the accepted procedure in those days.

Unfortunately, the system was filled with an explosive mixture of hydrogen and air when he found the location of the leak.

The resulting explosion completely disintegrated the entire vacuum system. The student then unilaterally decided that his thesis experiments were complete by definition. He gathered up his notes and, locking the door behind him, never returned to the lab. (Incidentally, he did receive his doctorate and, I understand, later had an excellent career in industry.)

Meanwhile, my first task was to clean up the mess before beginning the construction of my own vacuum system. In the course of the clean up, I undoubtedly inhaled far more mercury vapor than the EPA would now allow without noticeable ill effects.

Being an alien at MIT in those years presented some additional challenges. President Roosevelt had caused a new set of immigration regulations to be issued. This enabled the Immigration Service officials to say "no" based either on the old or new regulations which were applied in parallel. I was once trapped in Montreal for three weeks, rather than the intended weekend, because I lacked a statement from the Cambridge police chief that I had not been arrested in the past year.

However, my principal concern was that Canada was at war and I had a growing feeling that I should be doing something about it. So I sought to accelerate the progress of my thesis. My work days became noon to four a.m. with a neglect of the bounty of art and music that Boston offered. It was hard work; I later resolved never to work that hard again and, with a few brief exceptions, I have honored my resolve.

Twenty-four months after I arrived at MIT, and twenty-four pounds lighter, I left with my doctorate in physical electronics to take up radar design at the National Research Council, Radio Branch, in Ottawa Canada.

A. James Ebel

I graduated from the University of Iowa in January of 1937 with a bachelor's degree. I decided to get an advanced degree in engineering. I selected Purdue University because I could get part-time work at WBAA, the Purdue University radio station. After moving to West Lafayette, enrolling in Purdue, and setting up housekeeping with my wife, a new daughter, and two university student renters, another opportunity came my way: the University of Illinois (U of I) needed a chief engineer for their radio station.

They (U of I) were planning to build a directional antenna at a new transmitter site and decided from reading my article in *Electronics* that I was the ideal man for the position. They insisted that I come to Champaign-Urbana, Illinois for an interview on the university campus. They asked me what it would take for me to come to Illinois. I made a "ridiculous" request because I really didn't want to move.

They accepted this "ridiculous" salary request. I found out later that I could have gotten two hundred and fifty dollars a month more if I had asked for it. Moving, after only one month at Purdue, was traumatic but well worth the effort.

Radio station WILL, at the University of Illinois, was a 1000-watt daytime station with a "T" antenna between two towers on the campus. The transmitter was in the studio building.

Construction had already begun on the two-tower directional antenna system which had been designed by Jansky and Bailey. It was necessary to design the antenna phasing system, the feeder system to the towers, and the tuning units at the base of each tower. It was also necessary to buy a new 5000-watt transmitter, which was a RCA 5-D High Level Modulated transmitter.

When it came time to tune up the antenna system, I ran into problems. I had never tuned one up before, my antenna work had all been theoretical up to that point. I didn't realize the havoc mutual impedance between two towers could wreak in the tuning process. When you tuned one tower to get the impedance matched, you tuned that tower alone. And then when you tuned the other tower alone, it seemed that everything would be all set—not so—the mutual impedance between the two towers threw everything off. Any change while tuning one tower affected the tuning of the other. It was like trying to pick up a glob of mercury. After hours of sneaking up on the calculated values little by little at both towers, the design operating parameters were achieved.

The null measurements were surprisingly close even though the pattern didn't have a deep null. Field measurements for "proof of performance" worked out satisfactorily, largely due to the experience I had gained while working with Glen Gillette.

In the fall of 1938, I enrolled for graduate study at the University of Illinois to obtain a Master of Science in electrical engineering. As I expected, a lot of under-graduate engineering courses had to be made up, such as the course in electrical machines (motors and generators); a course on AC measurements; and a number of nonelectronic courses. I was also in charge of the engineering operations of WILL, so the number of courses I could take per semester were limited. I finally received my masters in October, 1943.

In the operation of WILL, we used male students on a part-time basis to handle the duties of our engineering staff. When World War II broke out in 1942, it became obvious that there was going to be a need for many young men to be trained in military electronics. For the operation of radio station, WILL, practically the entire staff had to either be women or men (4-F in the draft).

The women learned control room operations easily. And because of a natural manual dexterity, they were able to outperform some of the men we had when it came to intricate switching, record playing, and gain control operation. We were fortunate to find several women with First Class Operator's Licenses, and were able to train several more—so the operation at WILL went forward unimpaired during the war.

In 1933, Colonel Edwin H. Armstrong had developed wide-band frequency modulation for broadcasting. I heard Colonel Armstrong explain the development of FM at an I.R.E. convention, and also observed a demonstration of the fantastic performance of this system—eliminating noise and providing high fidelity audio transmission.

The Zenith Radio Corporation had one of the first FM radio stations operating in Chicago. So, in order to receive this signal, I built a Yagi antenna on top of the auditorium on the campus of the University of Illinois. For those who wanted to hear this fantastic new type of radio broadcasting, a trip up a narrow stairway to the attic of the auditorium was necessary.

The station had an irregular schedule because it was used mostly for demonstration purposes. After receiving my Master of Science in Electrical Engineering in 1943, I was appointed Assistant Professor of Electrical Engineering, teaching one course on the theory of thermionic vacuum tubes. (They don't use those much any more, do they?) I also started a consulting business on the side, filing applications for a number of FM and AM stations in the midwest, and designing new studios for several stations.

Emil Gaynor

Unfortunately, while I was cultivating this varied interest in scientific and technical things, I neglected my other studies which I found quite boring. As a result, my mother was called to the school on several occasions to hear how bad a student I was. My parents, however, were very supportive and in spite of severe economic problems managed to buy me a microscope and, more importantly, a super regenerative radio receiver kit.

I decided to go to Stuyvesant High School (a technical high school) so I had to take an entrance exam. My homeroom teacher told me I wouldn't make it and not to go. I ignored her, played hookey, and took and passed the exam.

The school in downtown New York had a superb and dedicated faculty and oustanding labs. One of the teachers who became a lifelong mentor and friend ran an after school lab in physics and electronics. We spent our afternoons doing some very fascinating things .

As I approached graduation, I attended a session sponsored by the Polytechnic Institute of Brooklyn to attract students. They held an essay contest to describe why we wanted to be engineers. I entered and won a slide rule that is a keepsake. Not being able to pay the tuition, I used the excellent machine shop training to get a job in a tool room. I enrolled in Poly's night school.

Later I went to work for the old Federal Telephone and Radio Company in New Jersey. This resulted in a very long and time consuming commute that allowed me to catch up on homework and reading. The U.S. was now involved in World War II. I was offered a job as a tool designer with the Consolidated Aircraft Co. (now General Dynamics) in San Diego, CA. Knowing I would soon be drafted, I went to the west coast for about ten months before Uncle Sam called me.

Enrico Levi

My engineering studies at the Politecnico di Milano in Italy were interrupted when World War II began. I had to transfer to the Technion, now the Israel Institute of Technology. There I had the privilege of studying with such teachers as Ollendorf, Naot, and Kurrein. In 1941, I received my first academic degrees, BSc's in mechanical and electrical engineering, and got married. All my subsequent studies were made with the encouragement—and financial support—of my wife, Nechama Bitia. In 1942, I received another degree fashioned after the German tradition and called "Diplom-Ingenieur."

My first job was in Haifa Harbour. Since I was not British, I was hired and treated with contempt as a "native"; the official classification was "unskilled labor," and the salary was commensurate. In practice, my job was to restore completely the electrical systems of ships that had been sunk and then refloated. This included desalting, redesign, and rewinding of motors and generators of every type and size. Another task was to fit Liberty ships that, more often than not, were put to sea in a great hurry and without many essential components. It was a lot of fun and a great opportunity for getting field experience.

In September, 1955 I enrolled as a graduate student at the Polytechnic University, at the time called Polytechnic Institute of Brooklyn. To this date, it has remained my homebase. In June, 1956, I got a master's degree in electrical engineering and a year later I completed my doctoral dissertation. The haste was due primarily to the inadequate financial support provided by a fellowship, and also to my advanced age—I was 38 years old!

It was my good fortune to have thought of a good idea, for which I later got a patent. Unfortunately, the device was a magnetic amplifier. (This recently has regained popularity under the name of "magnetic switch.") However, it was no match for its solid-state counterpart then being developed.

Thelma Estrin

Because of WWII (1941-1945), I left college during my freshman year to take a three month war training course at the Stevens Institute of Technology in New Jersey. Upon completion, I became an engineering assistant at the Radio Receptor Company in New York City. I worked in the laboratory tool and model shop where I was its only "lady machine operator." This experience led me to start my studies in engineering.

I also married Jerry Estrin, a history major, who then enlisted in the Signal Corps. After the war, we decided to leave New York and go to the University of Wisconsin. Here we both switched to majors in electrical engineering.

At the university, none of my classmates took me very seriously. Most thought that Jerry was keeping me in school to "keep me out of mischief," or to help him with his homework. But I took myself seriously and so did Jerry.

I recall that when I was a junior, and received the highest grade on an electronic quiz, the professor announced to the class that I would now be able to fix an iron when I was pressing Jerry's shirts! Another professor, who I had for DC and AC machinery, was pleased with my success but made me promise that I would be sure to have at least two children, to keep up the birth rate of well educated people. (We do have three daughters and two are members of IEEE.)

As a graduate student, I could not get a research fellowship, because the feeling was that I would not adequately use the scholarship to pursue a lifetime career. Instead, I became a teaching assistant. This delayed my Ph.D. for about a year. Fortunately, the professor, with whom I received my Ph.D., was very supportive. To this day, he still sends me all sorts of articles on brain research (which was my research area for about three decades) and now about women in the sciences, because I have become interested in feminism.

Harold W. Lord

My high-school subjects included general science and physics, the latter during my senior year. In connection with the physics course, there was a physics laboratory class which included a number of experiments. I finished ahead of the rest of the class, so my teacher, an older man, asked me if there was some experiment I might like to do.

I had noticed a small x-ray tube in one of the glass door cabinets. I asked if I could try to operate it. He readily agreed. I brought in my Model T Ford spark coil and toy transformer, and hooked up the high-voltage output to the x-ray tube. The tube was rather gassy, so it glowed visibly when excited by the spark coil.

I had borrowed a hand-held fluoroscope from our family doctor, and x-rays from the tube were made visible by its screen. When one's hand was held between the x-ray tube and the fluoroscope screen, one could see the bones of the hand. Each student of the class took turns to observe the bones of their hands.

In those days, most shoe stores had a small x-ray machine and fluoroscope set-up to observe the effect of the fit of a new pair of shoes upon the bones of one's feet. For this reason, there was no concern about the possible ill effects of x-rays upon one's health. I am still living, at the age of eighty-three, so apparently there was no ill effects from my brief experiments with this small x-ray tube.

John F. Bell

After two years, I had managed to save seventy dollars. Based on my radio service experience, I wanted to attend Purdue's electrical engineering school and become qualified for electronic engineering. It was an open question whether it would be possible to survive at Purdue with such a scanty start, but I did not want to delay any longer. I had a scholarship to enter Depaw University, but I found that I could go to Purdue without the scholarship cheaper than I could go to Depaw with it, so there was no point in considering Depaw.

A friend and I went to Purdue to enter in the fall of 1934. We talked to the Dean of Men, Dean Knapp, and, because the other boy was better financed, he encouraged him, but could see little hope for me. I then talked to Dean Potter, the Dean of Engineering, and I soon learned what a marvelous, perceptive person he was. I explained my lack of finances, my interest in electrical engineering, and that I had a scholarship to Depaw but I wanted to go to Purdue. I would need additional income to survive. He told me that a new program had just become available and he took me back to Dean Knapp for information about it.

It turned out to be a program sponsored by the Federal National Youth Administration (NYA) to aid needy students. The NYA enabled me to work at the university for thirty cents an hour. I asked for electrical work and was assigned to the telephone switchboard in the new women's residence hall with one hundred and thirty co-eds. A very fascinating position, but my schedule gave me no time to take advantage of it.

As it worked out, this just enabled me to make expenses *if* I went home on weekends and did the service work that had accumulated during the week, and *if* I made distinguished student status and got the thirty dollar refund on my fees each semester. This all worked out very well. However, there was nothing left over for safety or recreation, and time for study was very limited.

Needless to say, many times it was an open question whether I could continue. But in my junior, senior and graduate years I worked for the Purdue Research Foundation under Professor R. H. George, the developer of the pioneering Purdue Television System. This was much more interesting than the NYA university chores.

Professor George had finished a contract with a power company to study the transients on electrical power systems that were so destructive to power system equipment. In the course of this study, he developed the cathode ray tubes which he later used in his sixty line television system. The cathode ray tube was used in the first all-electronic TV receiver as the display device.

George Platts

The decision to enter engineering college came as a result of a most interesting experience at Walnut Hills High School. I was fortunate enough to enter with the first class in the old Walnut Hills High School building. This structure had been condemned years earlier by the city building department. However, it had been renovated to accommodate the first college preparatory classical high school in Cincinnati, and its first six-year high school.

One year, an assembly of all the students was held in the room normally used as a study hall. The purpose was to allow some senior students to demonstrate radio equipment which they owned. It was fascinating to me and I resolved then and there to study "radio" engineering in college.

Of course, there was no radio engineering degree offered at the University of Cincinnati (UC). As a result, I registered for an electrical engineering degree. The co-op engineering course originated by Dean Herman Schneider was a godsend because it enabled me to buy tuition, books and fees with my job earnings. I also had the advantage of living at home, thus avoiding room and board costs. I supplemented radio engineering by building radio sets at home, starting with a crystal set and finally graduating into more sophisticated ware.

The radio courses available were meager, indeed, but we did get a few. As to the classmates' names, I can recall a few: Henry Suter, who had worked at Bell Telephone Laboratories for a few years before entering UC; Paul Goodell, who could not finish because his co-op job ended and he needed the income to pay his way; Ben Ross, who died a few years after graduation; Frank Fugman; and, I particularly remember my alternate (on the co-op job), George Pettibone. He did me a great favor by letting me work in his place during the six weeks vacation period when he wanted to go home to Rockford, Illinois. It seems I only remember a few names but, after all, we only had sixteen students in our graduating class. Of course, the class of alternates we never saw.

I have always had difficulty with people at UC in trying to explain that the electrical engineering degree we got was not a bachelor's in electrical engineering. It was explained to us—at the time of the award—that the combination of academic studies and practical work elevated our status to electrical engineer. If you consult the records, you will find that the co-op jobs we held had to be approved by the coordinator to be certain the proper practical experience was being obtained on the job.

I remember one class member had a well paying job when registering at UC. The coordinator would not approve it, so he had to take a lesser job to qualify. I worked the five years with the Cincinnati Bell Telephone Company with continuous rotation to different departments.

Hyman Olken

I graduated from Harvard in 1929. I spent four years there in the radio engineering course, then went back in 1930 for a master's degree.

At that time, Harvard was one of the only two engineering schools that gave a course devoted entirely to radio (electronics) engineering. All other engineering schools gave four-year electrical engineering courses in electrical power engineering which included one or two elective one-year courses in radio engineering—usually a course in amplifiers or oscillators.

A few years before I got to the Harvard Engineering School, it had been a graduate school. Then along came a shoe manufacturer with a lot of money to give away (Mr. McKay). He offered to give Harvard twenty million dollars for its engineering school: provided they made it an undergraduate school. So they took his money and let students in directly from high school.

However, Harvard did not change the curriculum. The courses were all still graduate level, and we undergraduates sweated it out. We got an advanced technical education, but the going was rough.

Another interesting feature was that all courses in the engineering school were also available as courses in physics in the physics department of the college. So I enrolled in the physics department of the college, but took all the courses listed in the radio engineering curriculum at the engineering school.

This left me, at the end of four years, with a few of the most advanced courses in radio engineering program still to go. I took them, also as physics in the college, and received a master's degree upon completing them.

Robert McLane

My father, of whom I have only the fondest memories, highest respect, and eternal gratitude, said to me, "You are not meant to follow in my footsteps. Go to college and prepare yourself for something better."

Prior to WWI, my father had been employed by the CB&QRR (Chicago, Burlington and Quincy Railroad) in Burlington, Iowa. His older brother was married and had a son. My uncle was not drafted, unlike my Dad. During WWI, this brother advanced in the offices of the CB&QRR. When Dad was discharged, "nepotism" was a real entity in the workplace and he couldn't return to the railroad.

Thus, Dad started his own letter shop. His shop was involved in mimeographing, multigraphing, and mailing services. The business eventually grew to include printing "IBM-style" tabulating cards for the first computers of IBM's infancy. Dad said, when I returned from WWII, "Now, you have the gift of the GI Bill. Use it. Choose a career and get a rewarding job."

Chapter three

War stories

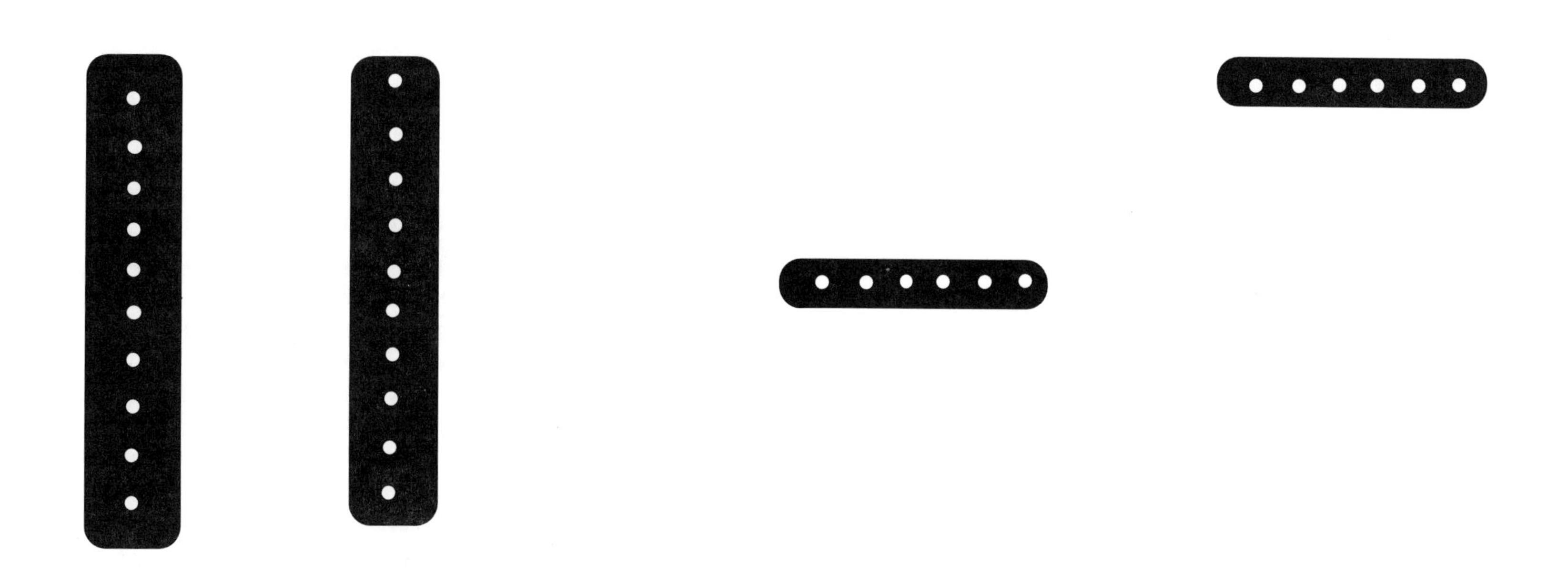

Paul D. Andrews

I enlisted in the Navy at age eighteen in 1918. The Radio Officer at the League Island Navy Yard gave me a brief oral exam and I was given a rating as 3rd Class Electrician (RO).

I received my basic training at Cape May, New Jersey. After a brief stay at the Navy's Cherry Head Radio school in Philadelphia, I was sent to the Navy Radio School at Harvard University along with my friend, Orrin E. Dunlap. Our training included the "High Power Course" which we thought would get us assigned to the Navy's station in France. However, we were both assigned to the new Navy station NBD in Bar Harbor, Maine.

The principal problem with communications was "static." This led the Navy to put the station as far north as they could. A wealthy Italian named Fabbri owned the estate on which the station was located. He was a radio amateur and welcomed the Navy's presence. The Navy made him a Navy commander and put him in charge of the VLF station.

The winter of 1918-1919 came before there was time to complete the buildings. A few small receiving shacks were built, one for each of the VLF stations. These stations copied everything transmitted by the British, French and German stations.

It was believed that electric wiring brought in static. Thus, no electricity was permitted in the receiving shacks. The only light at night was by kerosene lantern and the heat was provided by kerosene heaters. The fumes on sub-zero nights were terrible. As the barracks had not been built, everyone but the officers lived in tents heated by kerosene stoves. The mess hall had been started but had no roof so you drank your coffee fast before it froze

We stood watches of six hours on and twelve hours off. Most of the transmissions were in five-letter code at about twenty words per minute which was all we could handle. One station had a machine sender at about fifty words per minute. To handle this, we used a oscillograph-like photographic recorder made by the General Electric Research Laboratory. We had one operator who could copy fifty words a minute so when the recorder failed, the call went out for Benny Suter.

The VLF station was known as "Otter Cliffs." On the other side of the island, the Navy built a transmitting station known as "Sea Wall" which was equipped with a standard Navy shore station five kilowatt spark transmitter working into a 12 wire flat top antenna with a 12 wire down-lead supported by wooden towers, three hundred to three hundred and fifty feet tall. The radio shack was built almost under the antenna and we lived in a deserted farmhouse nearby. From this station we handled contacts with the Navy seaplanes on their history making flights across the Atlantic. We also handled the contacts with President Woodrow Wilson's ship when he went to Europe at the end of the war, as well as thousands of contacts from the returning military forces and their families.

At the end of WWI, the Navy was short of radio operators. It was only with the help of our friendly congressman that I finally was released from active duty in early September, 1919.

Harold W. Lord

Within only a day or two after the Pearl Harbor attack (Dec. 7, 1941), an electronics laboratory was organized under the direction of my former manager, Mr. W. C. White and his assistant, Mr. E. D. McArthur. Mr. White asked me to come back and work for him in this new laboratory, which was charged with the development of microwave magnetrons, and associated transmitting and receiving circuits for radar systems. I gladly agreed to do so, and was the first employee, after White and McArthur, of that laboratory.

There was to be a strong relationship for the duration of the war, among the MIT Radiation Laboratory and several large companies, which included the General Electric Company, the Westinghouse Company, the Bell Telephone Laboratories, the Hazeltine Corporation, and a number of others. To bring me up to date in the radar field, I spent two weeks at the Radiation Laboratory learning some methods being used to supply pulses of power to the microwave magnetron transmitter tubes as well as some of the lower-frequency older systems, which preceded the development of the microwave magnetron.

While talking with a Radiation Lab engineer concerned with supplying power pulses to magnetrons, he cited the need for transformers which would permit sending pulse power over 50 ohm impedance cables. This would eliminate the need to have the whole pulse-power generating system located close to the microwave transmitting magnetron. This problem intrigued me, so when I got back to my job, I reported this to Mr. White and he gave me the "go-ahead" to work on this problem.

Some of my theoretical physics friends thought I was foolish to attempt to design and build such a transformer. To transmit a one-microsecond pulse with a rise-time of 0.1 microsecond rise-time and 1000 pulses per second, requires a theoretical band-width of lKHz to 20MHz.

I designed a transformer to operate at 12KV 12 amps. l.0 micro-sec. 1000 per sec. pulses on the high-voltage winding, and a secondary of 4.5KV and 57 amps pulse. A pair of these were built and taken over to the Radiation Lab to try in one of their pulse transmitter systems.

They were tested by connecting one unit as a stepdown transformer between the pulse power generator and a 50 ohm cable, and the other unit between the other end of about thirty feet of the cable and the magnetron micro-wave transmitter tube. The system operated satisfactorily during a brief test, but I observed some severe corona between the high-voltage windings and the low-voltage windings.

Encouraged by these results, I sought help from our insulation specialist in the General Electric Research Laboratory. He was working on the development of a solventless varnish which, in the original liquid state, could impregnate a paper insulation system and then when heated would polymerize into a void-free solid state. This made possible the production of dry-type, corona-free pulse transformers for operation up to 12 kv pulse voltages. Oil filled pulse transformers were also developed by General Electric and Bell Telephone Laboratories. Westinghouse developed a different type of void-free insulation system for their power pulse transformers.

I devised a winding configuration suitable for power pulse transformers of step-up ratios between three-to-one and about seven-to-one.

This provided a bifilar type of secondary to supply power to the magnetron filament, and which, for a given power and pulse-width rating, had a lower L C product than all other coil arrangements. After the war was over, a patent was issued covering the winding configuration, which had become known by pulse-transformer designers as the "Lord Winding."

The work on pulse transformers was also supported by the development of thin lamination steel such as oriented silicon steel and of alloy steels such as Molybdenum, nickel, and iron (Monimax) and silicon, nickel, and iron (Sinimax). Various punchings of such thin materials, down to .002" thick were developed for pulse transformer uses, such as V I, and D V (overlapping V) laminations.

General Electric Research Laboratory cooperated with me in those developments and made temporary dies to punch these thin gauge materials for developmental pulse transformers. The Allegheny Ludlum Steel Corp. also cooperated in this effort when such laminations were needed to produce pulse transformers in large quantities.

About a year and a half before the war ended, I was "loaned" to the Naval Research Laboratory (near Washington, DC) to participate in the development of the Mark V IFF (Identify Friend or Foe) system. My responsibility was to design and develop all of the transformers required by the new system and by all related and required test equipment. This involved small power transformers for operation over the range of 25Hz to 1000Hz for test gear, and pulse transformers for handling single as well as a series of pulses as required by the coding equipment. General Electric set up a unit in their Specialty Transformer Department. in Fort Wayne to produce the prototypes of my designs.

In this assignment, I worked Monday through Friday at NRL, took a four p.m. train to New York City and from there a late night train which got me into Schenectady about seven a.m. My wife met me and took me home for breakfast. I then worked all day Saturday at my regular job there. (I had three assistants to oversee.) Sunday evening I took a six p.m. train back to New York City and from there a night train and so-called "sleeper" which arrived in Washington at eight a.m. When the work so required, I would take an over-night train to spend a day or two at the Fort Wayne GE plant or again an over-night trip to Boston and the MIT Radiation Lab.

I kept a record of my travels and during the first six months of this assignment, in addition to working six days per week, I traveled over twenty-six thousand miles by train! When the war ended, my work at NRL terminated.

Rowland Medler

The war intervened and I took two moonlighting jobs; teaching basic electronics at a local trades college and teaching an electronics course under the Engineering Science and Management War Training Act at what is now the East Tennessee State University.

When my draft number came up, the military drafted me, the teacher, although my students had been deferred for the thirty-nine week course. In disgust I volunteered to the Navy and promptly flunked the physical exam due to an old hernia suffered in the cotton mill.

The draft board was standing on the depot platform when the train returned me home. I was promptly hustled off screaming and kicking to Chattanooga to be inducted.

Due to unconventional answers regarding questions and demands for swearing allegiance to anyone wearing a tin sign on his collar, I was classified as a conscientious objector. I was now hustled off to medical basic training in Little Rock, Arkansas. I couldn't then and still can't pitch a sore thumb. However, I still did go through five battle zones in the Pacific as a Private First Class (P.F.C.) in reprisal for my stubborn attitude. That was long before it was popular to conscientiously object to war, but in the same circumstances, I'd do it again.

However, this wasn't as bleak an assignment as it could have been. It seems the military had ways of talking out of both sides of their mouths. I was put on temporary duty to the Signal Corps and assigned to a group setting up signal supply depots all up the chain of Pacific Islands.

Since this was considered combat duty, they shuffled my papers to a Special Services outfit. I turned out, as a P.F.C., to be in charge of depot and repair shop facilities staffed by those up through full Colonel. During "spare time," I built the first of a chain of broadcast stations in the islands from battle fatigue junk. This grew into the Armed Forces Radio Service. Upon Japanese capitulation, I applied for rescindment of the Commanding Officer status and made five advancements in rank in three days. This was also infraction of Army S.O.P (Standard Operating Procedure).

The guideline for these papers suggests we include our biggest goof. Mine is easy to define. On return from the Army, I opted not to accept GI Bill education. I was so violently sick of the military I didn't want to ever see another insignia, much less sign my name to papers containing Army logo.

In retrospect that was a mistake, but in the same circumstances I'd do it again. And the outcome was good in the long run. It let me pick up my job where I left off and even better, at thirty-three years old, marry the girl of my dreams and start my family. Forty-one years later, I wouldn't swap my happy home and three beautiful daughters for all the GI Bills on earth. However, it's a pity that both courses weren't possible simultaneously .

John F. Bell

Another interesting project was the 602 radar. This had a six foot parabolic reflector. One day the purchasing agent was shown how the system worked. He asked, "Do you mean to tell me that the signal goes out two hundred miles to an airplane, bounces back and goes into that tiny one inch hole in the center of the parabolic reflector?" We had to tell him that is exactly what happens—not much of the signal, but enough to amplify and display on the radar screen.

George Platts

Time marches on! In the fall of 1939, Hitler invaded Poland. There was no doubt what his eventual aim was. President Roosevelt knew that the U.S. would have to enter the war. Remember the Lend-Lease Program? Remember the destroyers we sent to Great

Britain? I was a member of the Naval Reserve. On June 9, 1941, almost ten years after I had graduated, and almost six moths before Pearl Harbor, I was called to active duty as a Lieutenant — Junior Grade, United States Naval Reserve.

Strangely enough, I was ordered to the office of the Inspector of Naval Material. Immediately after Pearl Harbor, The Crosley Corporation was given a contract to build the proximity fuze. The Commanding Officer of the Inspector of Naval Material called me into his office and showed me a crude sample of what turned out to be a proximity fuze.

He said, "Did you ever see anything like this?"

I replied, "No, sir, but it has a clear plastic nose in which there appear to be some small vacuum tubes like those used in a hearing aid."

He then said, "Crosley is to make five thousand per day, but first they must develop a practical manufacturing model to put in production. I want you to take charge of the office at the Crosley factory. Their contract is to be of the cost-plus-fixed-fee type and will require close management by your office."

"But," I said, "I was working for Crosley when I was called to active duty. Would the Navy approve of that?"

He replied, "You are the only electronics engineer I have, and we cannot delay."

So, we worked day and night with the Crosley people and representatives of the Applied Physics Laboratory which had developed the crude model. In one year's time, five thousand per-day were being produced.

What was the proximity fuze? It was a device which, when incorporated in an anti-aircraft shell, would explode if it came within range of a plane. Formerly, all the Navy had was a time fuze which would strike the target only by accident. I went on to open another plant in the east to expand Navy production, another to make fuzes for the British Navy and, finally, one to make fuzes for the Army.

When the war ended, I wondered how I would find something else this exciting. Obviously, I had moved from an engineer status to a manager status, but it was still a challenge to find something equally exciting. Well, I made three stabs at it and, lo and behold, I entered the General Electric Company almost exactly twenty years after they made me their original offer.

Philip Morris

Then the war came and I started looking for a spot in the military where my talents could be best utilized. I found an Air Force meteorology training program that was ideal. Normally the Air Force accepted men with two years of college and then sent them to meteorology cadet school. Upon completion of the cadet training, the student was a meteorologist with the rank of second lieutenant.

With the war, the Air Force figured that they would need a much larger number of weathermen than could be turned out with the existing arrangement. The Air Force set up a premeteorology training program for men with a high school education and another program for men with one year of college. These premeteorology students would be sent to colleges for intensive studies to bring them up to the two years of college level. Then they would go to the cadet school. I signed up with alacrity.

After I had a short stay at Bowdin College and a longer stay at the University of Michigan, the Air Force discovered that the casualty rate for meteorologists was practically zero. All these premeteorology students were not needed.

We were offered a number of options one of which was the Army Specialized Training Program in engineering, medicine and dentistry. Aeronautical engineering was not offered so I signed up for mechanical engineering which was close to my interests. I was sent to the University of Nebraska where I was informed that I was going to take electrical engineering. Protesting that I did not want to be an electrical engineer, I was offered the option of returning to basic training. I became an EE student.

After nine months at the University of Nebraska as an electrical engineering student, the army decided that the ASTP was to be scrapped and I was sent off to Fort Monmouth. There I went through radio repair school, a basic radar course and then to the AAN/TRC-6. This was a six channel line of sight radio-telephone set using pulse time modulation (it was called pulse position modulation then) and operating around 5GHz. We were formed into a signal service company and half of the company got to Europe as the war ended there.

After a lot of training exercises in the woods of northern New Jersey, we were ready to go to the Pacific; however, the war had just ended there. We didn't have enough points to be discharged so the Signal Corps had us take various telephone courses.

Next they sent us to the Pentagon where we were to be trained on that super secret encoding equipment used for communications among the various army headquarters. (It used recordings of tube noise for the encoding.) After training we were to be sent overseas to replace technicians there who had enough points to be discharged. Before this could happen, the military declared that we had too many points to be sent overseas. So we were put to work pushing tape in the Pentagon Signal Center until it was time to be discharged.

Granville Porter

In late 1937, I enlisted in the United States Navy. After training in San Diego, I was assigned to the *USS Arizona* as an electrician's mate. In late 1939, I was transferred to the *USS Langley* for duty in the China Sea area.

My enlistment having expired in October '41, I returned to Pearl Harbor. There I became a civil service electrician installing early English radar on our ships. During the December attack, I assisted by fighting fires, carrying stretchers, and so forth from ships in the drydock adjoining my shop. Unknowingly at first, I had seen *Arizona* sunk in the first eight minutes after the attack with a loss of one thousand one hundred and seventeen of a crew of fifteen hundred. (One thousand one hundred and two are still entombed in the sunken hull.) The *Langley* was sunk while on a mission south of Java in February '42 with a loss of about eighty percent of her crew. I currently maintain contact with survivors of both ships through reunion associations.

Richard Schwartz

I greatly enjoyed my college program. Shortly after I started in 1940, the program changed drastically because of WWII. Almost everyone left on the campus in 1942 was in a military program. I graduated in three and a half years and was in active service two weeks later.

My Army Signal Corps career was undistinguished militarily, but it was of enormous benefit to me personally. I "grew up" in the service by coming into contact with people from walks of life other than mine. I learned about bureaucracy and how to get around it.

Samuel Seely

World War II meant my departure from City College of New York with a new career at the Radiation Laboratory at MIT beginning in 1941. During the next five years at Rad Lab, I was involved with the early studies on aspects of the magnetron, with the development of the SCR 542, a radar equipment of interest to the Seacoast Artillery, with the development of radar training equipment, with the assignment as head of a small group to work with the Australian Radiophysics Laboratory.

I sometimes attended the lectures by W.W. Hansen and was the principal editor of *The Hansen Notes* covering many details of the fundamental undergirding radar, including the theory of magnetrons, klystrons, microwave antennas, receivers, and waveguides. This set of notes (thousands of pages in all) covered two or three years. Upon the completion of WWII work at MIT, I resumed my academic career.

Jack Staller

In 1937, I enrolled in Northeastern University. After the first year, I chose mechanical engineering and received a BS in 1942 in the early days of WWII. My first job was with the War Department at the Springfield Armory doing development and testing on machine guns. My preference would have been a commission in the Navy, but my bad eyesight prevented that.

After about two years, the draft board sent me a notice to have a pre-induction physical even though I had a firm deferrment. After the physical, it was suggested that I take the Navy "Eddy" test for training as a Navy Electronic Technician's Mate. I took the test and had my deferment renewed and forgot about it. Then the Navy started bombarding me with letters saying, "You have passed the test—a great future in electronics awaits you."

WWII was a war people believed in and I decided to get into it. I resigned my job, got myself drafted and was sent to electronics school in Washington, DC, for a year with seven months at the Naval Research Laboratories. This led to field service in electronic installation and repair, mostly in Japan.

Julius Stratton

When MIT was asked in 1940 to establish the Radiation Laboratory as the center for radar research in the United States, Dr. Stratton joined the staff as a member of the Theory Group. He also worked on the development of LORAN (Long Range Navigation), which by the end of the war covered nearly a third of the globe with radio beams enabling airplanes and ships to determine their location.

In 1942, Dr. Stratton went to Washington as Expert Consultant to Secretary of War, Henry L. Stimson. When communications for ferrying planes across the North Atlantic proved unsatisfactory because of the proximity of the magnetic pole, he went to Labrador, Greenland, and Iceland to study the problem and subsequently recommended a very low-frequency system. In this post he served also as chairman of the committees to improve the effectiveness of all-weather flying systems and of ground radar, fire control, and radar bombing equipment. He visited Italy, North Africa and the United Kingdom to study radar bombing and to assist in planning the use of radar in the Normandy invasion. In 1946, he was awarded the Medal for Merit for his services.

The Radioation Laboratory impressively demonstrated the value of interdisciplinary research and, as the end of the war approached, Dr. Stratton and others sought a way in which its momentum and program methods could be sustained for peacetime research. This was effected through the establishment at MIT of a new Research Laboratory of Electronics, of which he became the first director. Its form of organization was so successful that it soon provided a pattern for interdisciplinary research in a variety of fields at the Institute. Its example was followed at other institutions as well.

Harry E. Stockman....(late 1930s)

Finishing college, I served as Associate Professor of Radio Engineering, and one of my obligations was the university short wave station, SM5SX. Around that time, my superior at the university, Professor Loefgren, was appointed by the Swedish government as a radio expert to serve in Finland. However, he could not accept because of ill health and urgent duties, so I was sent in his place.

This was quite an experience. Our little Swedish plane was shot at before we arrived at Helsinki, so we had an emergency landing in Abo (now Turko).

I was soon given a well guarded lab in Helsinki. Whenever the Finns shot down a Russian plane, the detonator was removed and the transceiver brought to me. My job was to get the circuit diagram, and other technical details, then send a report P.D.Q. to Headquarters, as well as to Sweden.

I found some very ingenious Russian circuits, unknown to me, at least. They certainly knew what they were doing in this "radio" field. They manufactured copies of the American metal tubes, all with Russian lettering, and in later measurements back in Sweden, I found that these tubes were just as good as their American originals.

Coming to the United States in 1940, I happened to run across an old newspaper showing a photo on the front page of a Russian spark transmitter, housed in a gigantic carriage from the previous century, and pulled by two pairs of oxen. The text under the photo read something like this, "That's all they got, spark transmitters, totally antiquated....."

The truth of the matter was that this transmitter, apparently captured by the Finns, was one of the jammers the Russians cleverly positioned along its border facing Helsinki. And do I have to tell anybody in radio or electronics how a good job of jamming they did?

Traveling in Finland during the war had its ups and downs. Occasionally, I had to go by bus, and as you entered the bus you were given a white sheet, unless you already had one. If there was an alarm (alarms were a daily nuisance), you got out of the bus as quickly as possible, swept the sheet around you, and dived into the nearest snowy ditch. Today, this might seem like a joke, but indeed it saved many lives.

William H. J. Kitchen

The Canadian people manufactured their own ships and manned them to form the third largest Navy afloat at the time (WWII). Ships were also built for United States, Russia, and England complete with radar echo sounding devices and anti submarine gear. Canadians built many of the required components as well.

However, at the outbreak of war, there was no electrical department in the Royal Canadian Navy. Electrical equipment—such as it was—came under the jurisdiction of the torpedo department.

With the advent of the "all electric ships," Captain Culwick, a professor of electricity at British Columbia University, headed up a new department of "green stripers." Selected special branch individuals were promoted and re-appointed to the Principal Electrical Engineer Overseers in the new "L" (electrical branch). They were distributed among the various dock yards in Canada to accommodate the growing complexity of the ships being built.

I was sent to the "Great Lakes" area which comprised ship and boat yards in Collingwood, Midland, Orillia, Owen Sound, Honey Harbour and Penetanguishine. Each of these builders was supplied with resident installation technicians the whole being supervised by the Principal Electrical Engineer Overseer.

My ship at that time comprised a Willys jeep painted Naval blue. I think I remember the number as RCN 36. It was a jolly ship, I mean jolly cold in the winter time.

At the termination of the great lakes shipping program, I was transferred to the east coast and appointed Principal Electrical Overseer of Shelburn Dockyard. On cessation of hostilities, I returned to air conditioning with the Crysler Air Temp stationed in Montreal. This arrangement did not work out very well since I do not speak French and my qualifications at the time did not allow me to practice as a civilian.

Edmond S. Klotz

I had completed two and a half years of a liberal arts education at Brooklyn College when I entered the Army in 1942. I, along with seven other trainees, was assigned to the advanced electrical engineering curriculum of the Army Specialized Training Program (ASTP). The seven other trainees were assigned to the Pratt Institute in Brooklyn, New York, while I, presumably because I came from Brooklyn, was assigned to Ohio State University in Columbus.

As luck would have it, the program at Pratt was shut down three months later because the Army needed infantrymen more than they needed future engineers. I remained at Ohio State for a full year during which time I attended classes for about forty hours each week.

Upon completion of the program, each graduate was interviewed by two civilians for a period lasting anywhere from thirty seconds to fifteen minutes. I received one of the thirty second interviews and concluded that whatever they were looking for they didn't find it in me.

Several days later, however, all of the graduates except me and one other received orders to proceed to Camp Crowder for radar training. My orders were to proceed to Knoxville, Tennessee. Upon arrival at the train station, I was to call a telephone number for further instructions. I wound up at Oak Ridge and, after about a week of interviews and killing time, I boarded a train for Lamy, New Mexico. Once there, I again called a phone number for further instructions. I spent the next two years at Los Alamos as a member of the Special Engineering Detachment assigned to the Manhattan District Corps of Engineers.

I worked inside what was known as the "Tech Area." I wore a white badge which indicated that I was a "scientist" as differentiated from those who wore other colored badges. The white badge entitled me to attend the weekly colloquia where I sat silently among the science world giants and didn't fully appreciate the significance of the event taking place.

My assignment was to design and test an antenna for "Fat Man," the name by which the Nagasaki bomb was known. I used a full-scale model of the gadget and made radiation pattern measurements across a canyon. The process was time consuming. Patterns were measured on a point by point basis using a remote source that was anything but stable and had to be hand tuned to vary frequency. Eventually, an acceptable design was completed and the war came to an end.

Early in 1946, the Los Alamos facility still needed technical personnel and it was difficult for Army enlisted men to obtain discharges unless they had been accepted at a university. I wanted to obtain a bona fide engineering degree and applied to Ohio State University where I had spent a year in the ASTP Program. The registrar would not accept me because I was a nonresident. The university was inundated with applications from returning Ohio GI's. I wrote to Professor Boone of the electrical engineering department who knew me from my ASTP days there. He interceded on my behalf because they desperately needed seniors to balance their program; most of the returning GI's were entering freshmen. Six months later I received a BEE degree.

Harold Alden Wheeler

Then came World War II (1941-45) and the military requirements superseded the broadcast manufacturing. Starting just before United States declared war, we undertook a few small projects for the Signal Corps, none of which went into service.

Then we were selected by a committee of NDRC (National Defense Research Council) to develop a "mine detector" for buried metallic mines. I was put in charge of this project and it became one of my specialties for a few years.

Under the committee's guidance, we built on the idea of an exploring coil that was used for "treasure finders." I invented a feature of concentric coils which made that idea practical. Our first complete design was flown to North Africa in time for the Allied invasion. It was used to clear buried mines ahead of a column of advancing tanks. Our design was manufactured by another company selected by the Signal Corps in a phony competition. It was designated the SCR-625 and hundreds were used by the

Allied forces.

At the same time, we were introduced to radar by the Signal Corps at Fort Monmouth, New Jersey. They engaged our company to design some equipment for a new system which was an adjunct to the latest tracking radar. It was the IFF (Identity Friend or Foe) invented in Great Britain. They had negotiated for U.S. to assume the responsibility for production design and large-scale manufacture. It was extremely urgent. We made preliminary designs for all components and delivered them to the Signal Corps. The Army and Navy "weren't speaking to each other" so the Navy came to us separately.

The Mark III IFF system comprised an "interrogator-responsor" (IR) associated with a tracking radar, and a "transponder" on every aircraft or ship that might be a target. The IR was an adjunct to the tracking radar. It transmitted and received a pulse in synchronism with the radar pulse. The transponder received this transmitted pulse and responded with a coded pulse that was displayed on the radar screen beside the pulse reflected from the target.

The British circuit for the transponder was an Armstrong superregenerative pulse receiver. This served also as a transmitter of the coded reply pulse. This pulse was displayed on the radar scope beside the echo pulse to identify a friendly target (so don't shoot). I had to develop the first complete theory of superregeneration, so we could make valid tests.

The Mark III IFF was in the highest frequency band we had experienced (VHF 157-187 MHz). This required new types of antennas and transmission-line circuits, so these also became my next specialty. One was a vertical monopole, dubbed the "life-saver" antenna in view of its wheel base, which was carried by every Allied surface vessel. It carried a tuned circuit in the base for impedance matching with a 50-ohm line over the wide frequency band. One "line stretcher" I designed was a strip between dielectric slabs between shield planes, a forerunner of the printed strip lines that were to become common after the war.

A by-product of the Mark III IFF was the first transponder beacon to be used as a landing aid for aircraft. In the air war over the Pacific, many of our aircraft had to land on the Aleutian Islands, which had very low visibility, so many were lost. We received an extremely urgent order to develop and deliver the "YH beacon" based on the Mark III transponder for use on landing strips on the Aleutians. The first complete unit was delivered with our design engineer for immediate installation. It revolutionized the practices of aircraft rendezvous and landing. We contracted for the manufacture and delivery of many units for installation at Army airfields.

M. Lloyd Bond

And then came THE WAR. On December 8, 1941, I volunteered to the Navy. Because radio (soon to be called "electronic") engineers were so scarce, I was an ensign! My work was almost entirely engineering. And, in my four plus years in the Navy, I was never aboard a ship (except for one three hour visit). My toughest challenge was setting up a specialized navigation radio equipment training school, and becoming its Superintendent of Training. Being quite inexperienced in management, I stepped on lots of toes and ruffled a lot of feathers. However, I gradually got promotions in spite of myself.

Emil C. Evancich

While a student at Kelly High School, I was drafted into the United States Army. I was trained as a medical technician in the 94th Medical Gas Treatment Battalion. We were sent to Europe and followed the 3rd Army through Europe. After the fighting ended in Europe, we were going to be sent to the Pacific but WWII ended before we got there.

After being discharged, I married an Army nurse moved to Chicago and attended the Illinois Institute of Technology. While at school, I worked at the Champion Air-compressor factory on an assembly line four hours a day.

C. Richard Ellis

I then went to work at the Naval Reseach Laboratory in Washington, DC, as a technician in the aircraft radio group. At the time, they were working primarily with British radar and IFF (Identity Friend or Foe) equipment. As the war continued, the equipment became U.S. made, and most of us technicians became Navy enlisted personnel.

Since no suitable Navy rate described our capabilities, we were given rates as "Specialist X." The booklet given to Navy enlisted personnel at discharge to describe their capabilities to prospective employers defined those rated as "Specialist X" as being primarily "pigeon trainers." Since I knew of nothing I could teach a pigeon, I never used my booklet.

Harry D. Young

In August (1945), Selective Service recalled me for another physical and this time I was accepted. I was inducted into the Air Force on September 18, ten days after VJ Day. There were tests to determine the best place for inductees. When interviewed, I expressed my interest in radar work but there were no openings. The interviewer suggested cryptography and I was accepted.

During my basic training, I was pulled out so the service could investigate my background for security clearance. After basic training was completed, I was held up awaiting an opening in the cryptography school. In the meantime, a team had been going around to various bases to test recruits for IBM Customer Engineer School. The IBM opportunity came through first and I was sent to Endicott, New York, for four months to be trained in maintaining various IBM punched card machines.

The IBM school was rather unusual. Army regulations required that we always wore uniforms. IBM insisted that we punch in on a time clock. The rational for this was the insistence that everyone, even the president of IBM, Thomas J. Watson, was on the clock. Classes started at eight a.m. and continued till five p.m. with an hour break for lunch. Our instructor was a former customer engineer from Texas who spoke in a dull, monotonous voice with no inflections. It was so bad that some of the students fell asleep in class. We were taught about the IBM punched card machines of that time. These included the high speed sorter, collator, accumulator/printer and keypunches.

The sorter could take a stack of cards and sort them into eleven stacks, one column at a time. The operator had to replenish the input stack as necessary and empty the output stacks when they became full. When the input stack was finished the operator took one of the sorted stacks and started again with the next column. Sorting a large data base was slow and tedious but considerably faster than doing it by hand. Sometimes the operator would drop the card stack as it was being moved, and the sorting process would have to be completely done all over again.

I went to Greensboro, North Carolina to wait for my assignment. There were many tales of men going through training and then never doing what they were trained for. From there, I went to New Orleans to board a troopship to the Panama Canal. I was assigned to Albrook Field in the Canal Zone.

The IBM installation had previously been in Hawaii. Panama generally used 25 cycle (now Hertz) power but the IBM equipment needed 60 cycle power. The Army ran a power line from another area of Panama for this purpose. When I plugged the keypunch machine in, I noticed an arc from the plug to the socket. There was quite a potential difference between the grounds. It did not seem to bother the equipment but the operators complained about the shocks. To solve the problem, I placed rubber mats under the machines and the operators' chairs to isolate them from ground.

Initally, the equipment was not operating so the IBM office in Panama City sent two men to assist. IBM employees have the reputation of always wearing dark suits with ties. One of the machines had an oil pan at the bottom to catch any oil drips. One of the men lay down on the floor in his suit and tried to get at something in the machine. As he was doing this, his hair dipped into the oil pan and he arose dripping with his new "oily do."

W. A. Dickinson

During the war, we made a variety of tubes for use in radar, loran, and so forth. Radar designs, including the display tubes, progressed rapidly with the whole industry cooperating and working with the Radiation Laboratory at MIT, the Navy Research Laboratory, and others. Our production reached twenty thousand tubes per month, all pretty much handmade, and at terrible scrap figures—thirty to forty percent as I recall. None the less, Sylvania was awarded the Navy "E" late in the war.

When the plant opened I had little concept of what we were getting into. Western Electric, which built radar, was my first visit to a customer. We had supplied them with sample five-inch oscilloscope tubes, which they had rejected. We didn't understand why and it was my job to find out what the problem was.

When I saw an RCA tube in the unit compared with ours, it was immediately obvious that RCA had made a new tube with a smaller spot size, which greatly improved the resolution and appearance of the display. From that day until my retirement, tube spot size and focus quality were the paramount concerns of my professional life.

A more successful project to which I contributed was a small (70 mm) tube for aiming radar in the British Spitfire. The display was very small—a bright spot on the tube face represented a target. The pilot maneuvered the aircraft to bring the spot to the center cross-hairs on the tube face, which indicated that the target was dead ahead, so the guns were aimed.

Our first requirement was that the tube be able to withstand severe shock and vibration. I designed a new U-shaped mount "snubber," which enabled the tube to pass shock and vibration specs. After we adapted one of our mount designs to meet the electrical specs, we received approval and promptly began filling orders for it.

We failed to file a timely patent application on the snubber, so it was never patented, but the Sylvania Parts Department sold millions of them. The principle of its design has been used widely since.

I (and many others) worked hard during the war—six day weeks with little time off. However, I didn't make any of the big money we heard about in other industries. The company kept me deferred from the draft, so I was practically frozen in my job. I regret that I never fought for my country, but I believe I contributed the most where I was.

I also worked in the Civil Defense Control Center and helped with Red Cross fund drives and blood banks, and bought savings bonds. Gasoline, sugar, shoes, and other commodities were rationed, but we "got by." Emporium was a good, safe place to spend the war. Also, it became known as Girls' Town (from the Colliers magazine article) for the three thousand or so (mostly young) women who worked in the tube plants. Maxine was one of them—we were married in May, 1945.

Max W. Kuypers

During high school I had no interest in engineering as my parents could not afford college. I just accepted that I would become an auto mechanic as I had mechanical aptitude. (I had learned about cars by keeping my "jalopy" running.) I had just gotten my first job in a garage when Holland was invaded by Germany.

I am Dutch born and was not an American citizen at the time (1941). I was called up for service by the Dutch government-in-exile in London. Not knowing better, I jumped at the chance and eventually arrived in London along with many other "volunteers" from all over the world. As with any military service, I was given an aptitude test; I was told I qualified for the Naval College to learn electrical engineering.

The curriculum was naturally oriented to shipboard electrical systems, with my last year specializing in radio/radar. Class work was tough and involved long hours. Lab work, as such, was usually on board a vessel in the yard for repairs.

I was commissioned in 1943, and as typical of the service, there was no immediate need for electrical engineers. When I graduated, I was posted to a merchant ship as the gunnery officer, for an eight man crew, to handle the four inch cannon mounted on the stern of the ship. I spent the war years as a gunnery officer and, while I had two ships torpedoed under me, I never sighted a U-Boat.

Edwin H. Miller

Tactical deception has been used for centuries to gain an advantage over the enemy. It can be used to cover friendly maneuvers, create illusions of strength or movement, gain time, or draw the enemy into a trap. Each of these can help the tactical commander achieve his objective if the technique is highly developed, timely and carefully coordinated with the real activity.

The 23rd Headquarters Special Troops, attached to the U.S. 12th Army Group

Headquarters, was a force of eleven hundred specially trained soldiers and, probably the largest organization of its kind ever developed. During the 1944-45 time period, the techniques and operations of the 23rd Headquarters were classified "Top Secret."

The personnel of the 23rd were carefully selected. For example, visual deception personnel came right out of Greenwich Village and other artist colonies around the country. Radio technicians were selected for operating techniques and knowledge of radio security (OPSEC). This would enable them to perceive how their phony traffic would sound to the listening enemy. My unit, the 3132nd Sonic Deception Unit, was mysteriously plucked out of Signal Corps schools, such as Fort Monmouth, New Jersey, and taken on a special train to Pine Camp in upper New York.

The various units trained separately in several locations in the United States. We did not participate in combined operations until we came together in the combat zones of Europe. My unit was not activated until March, 1944. We landed on Normandy's Omaha Beach after having only five short months to learn how to use the new equipment. Needless to say, the 3132nd benefited from a lot of on-the-job training.

The 23rd Special Troops utilized three basic deception techniques: visual, radio and sound. There were no set plans as to which techniques were to be employed. In fact, they changed from one mission to the other. Although born of the necessity to experiment, this probably helped to confuse the enemy and achieve the desired effect.

Visual. The principle visual deception equipment was the extensive set of inflatable facsimiles. These were particularly realistic when used in conjunction with camouflage nets or natural foliage. Though the Germans did not have much aerial reconnaissance, and the weather was rarely clear anyway, certain enemy sympathetic civilians were found to make good deception conduits provided they were allowed just enough visual contact. Artillery dummies were augmented at night with flash devices, and most of the time a few real vehicles were mixed in with dummies to add realism.

Many other forms of visual deception were used. Carefully marked phony command posts were guarded by pretend military police with fake shoulder patches and sitting in deceptively marked vehicles under an array of phony telephone lines. In direct response to a War Department observation that we seemed more like technicians than real fighting soldiers, men were commanded to drive through the countryside, goof off in all the bars and pay appropriate attention to the young women just like any respectable infantryman or armored soldier would. This was a tough assignment for a nineteen year old barely a year out of high school, but I managed. Even the officers got to promote themselves to general officer rank so civilians would believe that Division X was really in their neighborhood.

Radio. Radio deception techniques were usually the backbone of the deception "shows." This was primarily due to limited German air-borne reconnaissance and their tendency to believe in their own radio intercept technology and exploitation.

There were basically four types of missions for radio deception. Network substitution was used to cover the withdrawal of a Division from one place so they could surprise the enemy in another. Sometimes a green division was helped to take over from a seasoned one in a smooth transition which included the same frequencies, schedules, call signs, cover names, crypto keys, protocols and peculiarities. This is where the 23rd's radio expertise led to some real on-the-job training for the new substitute operators.

Other times the 23rd would imitate a radio network remaining silent elsewhere, or dummy radio traffic would enhance a visual and sonic buildup "show." The 23rd would also sometimes create the illusion of a network moving in one direction while it was really moving in another.

The 23rd's radio deception scope involved anywhere from ten to thirty radios and up to one hundred operators. On some missions, their field of operation covered one thousand square miles. The innovations included complete imitation of the covered unit through the use of nicknames, colloquialisms, venaculars and even regional accents.

Sound. Perhaps the most unique unit in the 23rd was the 3132nd Signal Service Company-Special. With the cover name Heater, we had the most technically sophisticated equipment in the 23rd and transported it in half tracked vehicles similar in external appearance to those of the tank destroyers. Under the canvas cover, behind the 50mm machine gun turrent, there was dual wire recorder/playback equipment whose output could be amplified to 500 watts of high fidelity audio sound via two 6 foot horn speakers or four 15 inch cone speakers. These rigs could project the sound of military activity up to six miles.

Heater had its own meteorological unit to measure wind direction, velocity and humidity for better positioning of the string of halftrack sound trucks. The sound scenarios were carefully scheduled. For example, we would create the sound of a column of tanks moving along a level road, down a hill, across a wooden bridge, onto cobblestone streets and into defilade positions. The coordination of these moving scenarios, which were usually done at night, was by field telephone wire to avoid "stage direction" over an interceptable radio network. Of course, the deceptive radio network coordinated the phony unit's movement in parallel with the sound and visual effects.

Operations. The 23rd conducted eleven missions between July 1 and December 31, 1944 and ten more in the first quarter of 1945. These missions ranged from three to ten days. Nine missions were considered successful, ten questionable and two unsuccessful. Seven of the missions utilized the full set of visual, radio and sound deception techniques. Eight used some combination of two techniques and six employed only a single discipline (visual, radio or sound). Of the nine missions considered successful, only three employed all the techniques, two used a combination, and four employed only one.

Mission success criteria was based on its contribution to tactical objectives, and not on whether the deception was successful per se. In OPERATION BREST, the deception was a success but its employment questionable. This was because a real armored unit tried to attack over a ridge line on which the 23rd had earlier created a phony armor attack build up. Apparently, there was a lack of communications between the 23rd and the real unit which allowed it to actually attack in an area which had been called to the enemy's attention.

The more successful operations were in the last three months that the 23rd was in combat. Most of these were based on better liaison with and cooperation from the unit being covered. For example, it did not do much good to try to cover a division's movement if that division did not move in darkness or try to conceal its identity on roadways.

It has frequently been asked whether the 23rd suffered considerable casualties, since it frequently worked within a few hundred yards of the enemy. Fortunately, the enemy didn't want to waste his ammunition or expose his position by firing at sounds in the night or hardly pinpointed radio transmitters. There was only one unfortunate instance. An unlikely harassing mortar round squarely hit a 6X6 truck killing the several men on board. Otherwise, we had no casualties except for a few non-fatal accidents.

Excerpted and adapted from "Tactical Deception in WWII," by Edwin H. Miller, *Journal of Electronic Defense*, October 1985.

Eugene W. Greenfield

In 1943, the war had come to our (United States) coasts. Isolated instances of U.S. ship sinkings at sea by German U-Boats had been reported from time to time. These were mostly cargo vessels carrying supplies and munitions to the European fighting fronts. As escorting by Naval vessels improved in experience and numbers, the at-sea losses decreased drastically. But then almost overnight, we started to lose oil tankers plying our coastal waters from Philadelphia up to Boston Harbor.

In none of these cases were any submarines involved. Thus, the only reasonable conclusion was that mines were responsible. These were obviously laid by German submarines in the waters of our busy coastwise shipping lanes and they were proving to be extremely effective. At one time, during this period, we were losing a tanker every day.

Of course, what mine sweeper vessels and gear the Navy had at that time were put into operation, notably around the ports of New York and Boston. These were conventional M.S. vessels using conventional mechanical sweeping gear to gently pick up and retrieve the usual floating or shallow tethered mine for later disposal. The sweepers were also prepared to use gun-fire to detonate these mines as circumstance permitted.

But no mines were found, even after painstaking coordinate search procedures with assistance from aircraft. This was a very perplexing situation. How then could we be having our ships blown up when there were no mines to be struck and detonated?

Two events, one off our coast and the other on an isolated English beach solved the mystery. The first event was the blowing up of a U.S. mine sweeper off the Boston harbor. This mine sweeper was a large double purpose ship able to destroy submarines as well as sweep for mines. It was equipped with very sensitive sound propagating and receiving gear (asdic) to detect location and distance to submerged submarines. Just before the ship was sunk, the asdic operator had picked up echoes from a small object submerged about seventy-five feet. The vessel was moving so slowly that the asdic operator could see the object range decreasing slowly for a moment before the explosion took place. Fortunately, most of the mine sweeper crew were rescued and uninjured—in particular, the asdic operator who was able to report his echo pickup.

In England, a large, steel ball shaped object had washed up on one of the barbed wire beaches. Suspecting that it was some type of mine, demolition experts were called to deactivate it. They had never seen anything like it. The object was slowly disassembled. Their wonder grew as the complexity of the electronic and magnetic detonating system for half a ton of explosives was revealed.

When the combined U.S. and British naval intelligence units had completed their study, it became apparent that this German mine required no direct contact with a ship for its detonation. It could be tethered as much as one hundred feet below the sea surface and still be effective. The initiation of the detonating mechanism was by a delicate magnetic device which when the mine was placed in position was balanced against the earth's magnetic field at that location. Passage of the steel hull of a ship would appreciably change the earth's magnetic field at the mine and thus unbalance its magnetic device. This set in motion the firing mechanism. It was also apparent that the magnetic unbalance had to proceed slowly from low to maximum, approximating a ship's passage, before it armed the detonating train.

At this time, I was a research engineer for a large wire and cable manufacturer. We were heavily involved in turning out enormous quantities of insulated cables for the Navy's ship degaussing program. In charge of the program for the Navy then was Captain Hyman Rickover. During the course of this degaussing program, I had frequent contacts with the Captain and his staff.

Experimental studies by the Navy had shown that a magnetic mine could be set off if a heavy, on-off pulsing electrical current flowed in its vicinity. To carry this out practically and with impunity, it was conceived that a mine sweeper would have a large d.c. generator with several thousand ampere capability and suitable on-off timed switching gear. The generator output would feed into a long, well insulated, trailing buoyant cable. At the end of the cable was a shorter bare conductor electrode putting current into the sea.

My company was given the problem of designing and manufacturing the buoyant cable and its electrode and that was the task I was directed to in all haste. We looked at a number of possibilities. The simplest was to have a large diameter water-tight hose on the outside of the necessary number and size of copper strands—laid helically—and over that, a thick rubber insulation and a strong jacket of asphalted braided jute. This simple design was the one adopted. The sealed end of the buoyant cable was fitted with a connector to which a flexible bare stranded cable could be attached to act as the sea electrode.

With Navy approval, we started manufacturing the cables. Samples, from the first few runs, were fitted up for field trials. For these tests, fifteen hundred feet of buoyant cable plus one hundred and fifty feet of bare electrode cable were wound on a huge reel mounted at the after end of the sweeper vessel As the ship got under way, the cable was unreeled and entered the water as a long trailing line. With the inboard end of the buoyant cable connected to the generator and the pulsing current started, any mine detonations would then occur at least a safe fifteen hundred feet astern of the vessel.

The tests were very successful and demonstrated that we could rid our shipping lanes of magnetic mines. However, there was one problem. When a mine was detonated, like as not the electrode section would be torn off of the buoyant cable, destroying the latter's end seal. When this occurred water would immediately enter the cable's hollow core and it would sink. Under that condition, the cable could not be retrieved and had to be abandoned—all fifteen hundred feet of it. Also, small craft inadvertently crossing over the long trailing cable could damage it sufficiently to open its protective covering. Again, the cable filled with water and sank.

My solution was simple but tricky to incorporate in the cable manufacturing process. We bulkheaded the cable. By extruding vertical sealing barriers every ten feet in its hollow core, the cable could sustain a number of severe damages and still remain afloat. We had to set up special x-ray equipment in the manufacturing line to visually check the barrier closures and make sure the bulkheading was effective. As we gained experience and developed improved extrusion techniques, making the buoyant cable became routine. We supplied the Navy with many hundreds of magnetic mine sweeping cable units.

With these new trailing cable devices, the Navy went to all wooden hull mine sweeper vessels designated YMSs. Using this precaution, a YMS had little effect on the earth's magnetic field and so had no need for a degaussing girdle. Another large improvement in sweeping efficiency came when two YMSs moving parallel to one another and separated by about twenty-five hundred feet, pulsed their sweeping currents alternately. For the pulse period when sweeper No. 1 was sending current into its electrode, sweeper No. 2's current was off. At the next pulse, sweeper No. 2 supplied the electrode current in the opposite direction while sweeper No. 1's current was off. This improved operating system did much to clear all of our coastal shipping lanes of magnetic mines for the rest of the war.

Some of my experiences working on this project were grim. For a day of sea trials, I reported to a Naval pier at the foot of Broadway in New York a little before six a.m. There the YMS was all ready to take off, just awaiting some Brass and a few engineering experts. By seven a.m. we were starting through the Narrows. When we met the ocean rollers, the little vessel pitched and yawed sickeningly, like a bronco. However, as soon as the long buoyant cable was unreeled into the water, it had a marvelous stabilizing effect. Then as we got under way along the prescribed sweeping course we might be so unfortunate as to pass through the ugly, sickening debris of one of our blown up ships. There was always a great oil slick and in it were all manner of floating pieces and garbage from the destroyed ship. We would slowly move around the area looking for bodies or possible survivors.

On the lighter side, I recall several occasions when Captain Rickover would go with us on these trial runs to check operation and management of the long, buoyant cable. As we took off and stood down the harbor, he was very much in evidence: checking generator output, how we were preparing the cable for reel-out, giving instructions and orders to us and the ship's officers and so forth. But as soon as the ocean swells started to lift and drop the vessel, Captain Rickover would disappear. I later learned he was subject to sea-sickness and preferred to wait out the roughness in his stateroom. There were rough sea times when I, too, would have preferred to be lying in bed, but no such luck for me.

I do want to say here that my admiration and esteem for Captain Rickover was very great. During the traumatic time of the magnetic mine sinkings, he organized the various teams who worked to eliminate the menace. It was his leadership, hard driving and never taking "no" or "cannot do"' for an answer that was largely responsible for our success.

Kenneth G. McKay

I became the project engineer for the Microwave Zone Position Indicator (MZPI). This mobile surveillance radar was suppose to operate at a wavelength of ten centimeters, and be sufficiently flexible to perform in the arctic or in the desert. It was impressive how much we accomplished so quickly with so few people; one person was responsible for the design of the receiver, another for the transmitter, a third for the antenna, and one for the display and controls. A mechanical engineer did all of the mechanical design. I did the "systems engineering" (if I had then known that term), a loose version of a Development Plan, and coped with the Canadian Army.

I was strongly supported by my immediate superior—Dr. D. W. R. MacKinley. Don was a thoughtful physicist who, post war, made some elegant studies of meteor trails using some of our radars. Our thinly manned MZPI project was characteristic of most of the Radio Branch projects and, more broadly, of many Canadian endeavors. It is extraordinary how rapidly a project can be completed provided: no basically new components need to be developed, and many conceivable failure modes are left unexplored, possibly to be uncovered later in the field.

In 1944, we transferred responsibility for the MZPI to Research Enterprises Ltd., a crown company in the Toronto suburb of Leaside. There the manufacturing would take place. I was the advisor to the Chief Engineer for manufacture. This entailed spending four days a week in Leaside, three days in Ottawa, and a year's worth of overnight upper berths. Here I learned about manufacturing. I had thought that our research prototype would, with minor modifications, be the actual model for manufacture. Instead, it was extensively redesigned—the art was moving quickly. Practical concerns outweighed theory. Under forced draft, production lines were rather rugged, but they did produce.

In the fall of 1945, the war in Europe was over but the British War Office decided that they still wanted the MZPI for postwar operations. (Indeed, it was later deployed extensively by NATO forces). We sent the production prototype to Malvern, England. We soon received a copy of a report that had already been widely circulated in the War Office. In effect, it said that the machine failed to meet most of the specifications; however, six months of redesign by British engineers could probably bring it up to scratch. Their attitude toward the colonies was evident.

I was sent over promptly to repair the damage—a mission that turned out to be one week technical and *two months* diplomatic. Actually, our British friends had not read the accompanying manual and had tried to operate the MZPI at ten percent below the minimum required voltage—half of the relays wouldn't close. With the proper voltage applied, I spent a week convincing myself that all of the specifications were indeed met. The following two months were spent convincing the author of the original report to join me in signing a second report that gave its complete blessing to the MZPI. Importantly, both reports had the same distribution list. The technical contributions of the British War Office had been to provide an extremely helpful supply sergeant.

Yardley Beers

MIT Radiation Laboratory—the war years: 1942-1946. At MIT, I was a member of the receiver group, which designed receivers for microwave radar. For a while I was concerned with developing low noise amplifier circuits. Later, I became involved in designing intermediate frequency amplifiers using vacuum tubes with "miniature" seven pin sockets and having single tuned interstage networks that were "stagger" tuned: that were not resonant to the same frequency. It had been found that stagger tuning produced a greater gain-bandwith product than synchronous tuning.

In this work, I applied a knowledge of theoretical physics to solve a practical production problem. One of difficulties was instability caused by positive feedback, which was worse with the miniature tubes than in earlier models using larger tubes. My colleague, P. R. Bell, had studied the causes of positive feedback and discovered cures for most of them.

One mechanism Bell postulated was due to the chassis acting as a wave guide beyond cut-off. For stability the wave guide attenuation per unit length had to be greater than the gain per unit length. Therefore the chassis should be narrow, and the tubes should be in a straight line. Although my amplifiers had this geometry, they were not stable.

I happened to attend a lecture given by Julian Schwinger (later Nobel Laureate in Physics) on the theory of fields in wave guides having obstructions like posts and irises. I recognized that this theory was mathematically similar to that of the Stark effect, the splitting of energy levels of atoms and molecules by an electric field. Then I recognized that the same mode functions that belong to a wave guide below cut-off apply to one beyond cut-off. I surmised that if I placed conducting posts across the middle of the chassis, with the correct orientation, the attenuation could be increased. Experiments supported my hypothesis, and my amplifiers became stable in mass production.
At the end of the war, I was assigned to be one of the many authors of the MIT Radiation Laboratory series of books summarizing the work of the laboratory.

Julius J. Hupert

I graduated from the Warsaw Polytechnic with a Dipl. Ing. degree (equivalent to MSc. in electrical engineering) as a pupil of Professor Janusz Groszkowski, then an internationally prominent contributor to the theory of oscillation. The country, too, was young with recently obtained independence. Professional prospects were good for a limited number of young, capable people. It was not likely that in more staid societies a person of my relative inexperience would be entrusted with the independent design of trans-atlantic transmitters and various ship transmitters for the Navy and Merchant Marine.

The institution itself, P.Z.T. (State Telecommunications Establishment), was an unusual European creation—state owned, but run entirely on industrial lines and responsible for its own profit, progress and maintance. The concept worked well.

In the face of German pressure in September, 1939, PZT ceased effective operation. A certain amount of radio equipment and its maintenance gear was transferred to a mobile Army convoy. A period of hide-and-seek with German tanks and bombers ensued until September 17 when, as a result of the Ribbentrop-Molotov secret pact, the Soviet army attacked from the other side. Squeezed between the two invading armies, with no armament of any kind (not counting the five bullets in my revolver), we decided to cross the Romanian border and somehow make our way to the allies in the West.

In one bout of vandalism, we destroyed the mountains of precious, pampered radio equipment, mounted our trucks and made a desperate dash to the southern Romanian port of Constanza. The Allied embassies, having been clandestinely contacted, were interested in our expertise. A curious tug-of-war started: the British and French embassies tried to bribe the Romanians with the purpose of getting us out to the West. The German embassy threatened and intimidated.

With a few friends, I managed to escape a not-too-tight internment, and obtained an "almost authentic" passport for exit to the West. In Paris, France, I was directed to Societe Francaise Thomson-Houston to carry on the design of a high-powered transmitter using the Doherty modulation system, then all the rage.

One morning (1940), we found everything disorganized in the face of a German approach. The army had previously retreated overnight. The railways being bombed, we secured five bicycles and started another episode of hide-and-seek. One of us did not know how to cycle—but one learns fast in such circumstances.

We decided to go to Havre on the chance of being closer to Britain and hopefully gain some information about our situation. We found the British forces gone, only one soldier and one airman left behind, drunk (supposedly left for this reason). We also found, totally by accident, a truck-full of Polish soldiers heading south, to be evacuated in St. Jean-de-Luz on the Spanish border. We sold the bicycles and joined them.

The next morning, in an inadequate anchorage and stormy seas, a rather dramatic embarkation took place. We sailed to an unknown destination in Britain. I recognized the port town as Liverpool.

In negotiations with Polish authorities, the British decided to make their own use of our talents, we being top experts in various fields. The initial selection was made by Polish authorities. The suggested candidates were then interviewed by the Director of Scientific Research of the Admiralty, who coordinated the scientific technological effort for all the armed forces.

I was selected and, after an ingenious interview, I was seconded to the H.M. Signal School (later the Admiralty Signal Establishment) in Portsmouth—then, on occasions, being heavily bombed. The entire establishment was later shifted to the quiet, small town of Haslemere in Surrey. The initial appointment took place in autumn of 1940. I was formally transferred to the Navy, and remained in the service of the Admiralty until my discharge and demobilization in 1947—two years after the infamous Yalta conference.

Another, much more extensive, activity of mine originated in a much less orthodox fashion. When at the beginning of my assignment to H. M. Signal School, I was given a tour of ships' transmitters of the then current design. I commented on their method of frequency generation, which required laborious setting by wavemeters and—in my view—could not guarantee adequate frequency stability anyway, especially in ship-to-fighter plane communication. The fighters had fixed-frequency crystal-controlled receivers. I was told that we could not use crystal-controlled tranmitters owing to the uncertainty of ship-operational assignments.

I then contrived a partial-crystal-control frequency generation scheme, flexible and yet requiring only a few crystals of fixed frequency. In a sense, this was an early precursor of modern frequency-synthesis schemes, although the term was not in use then. My scheme was reviewed and lauded as ingenious. But I was told that there was really no need to so extensively modify all the ships' transmitters dispersed all over the world. I was, however, permitted to initiate a small lunch-time project with the aid of one enthusiastic technician, provided the beacon-bouy project did not suffer.

Imagine, then, my feelings when one morning I was told the disastrous news of the aircraft carrier, *H.M.S. Ark Royal*, having been sunk in the Mediterranean. The tactical analysis of the events (I was not told the details) revealed lack of reliable raising of the fighters by radio.

Never before or since have I seen a project raised from a "lunch-time" to a "top-priority" status so rapidly! I was given access to all the facilities. The idea was embodied in a few design versions suiting various exisiting—and later (with a few ameliorations) future—transmitters. When, in 1947, I was leaving for the U.S., the post-war transmitters were being designed, using many of my earlier ideas.

Chapter four

Getting started

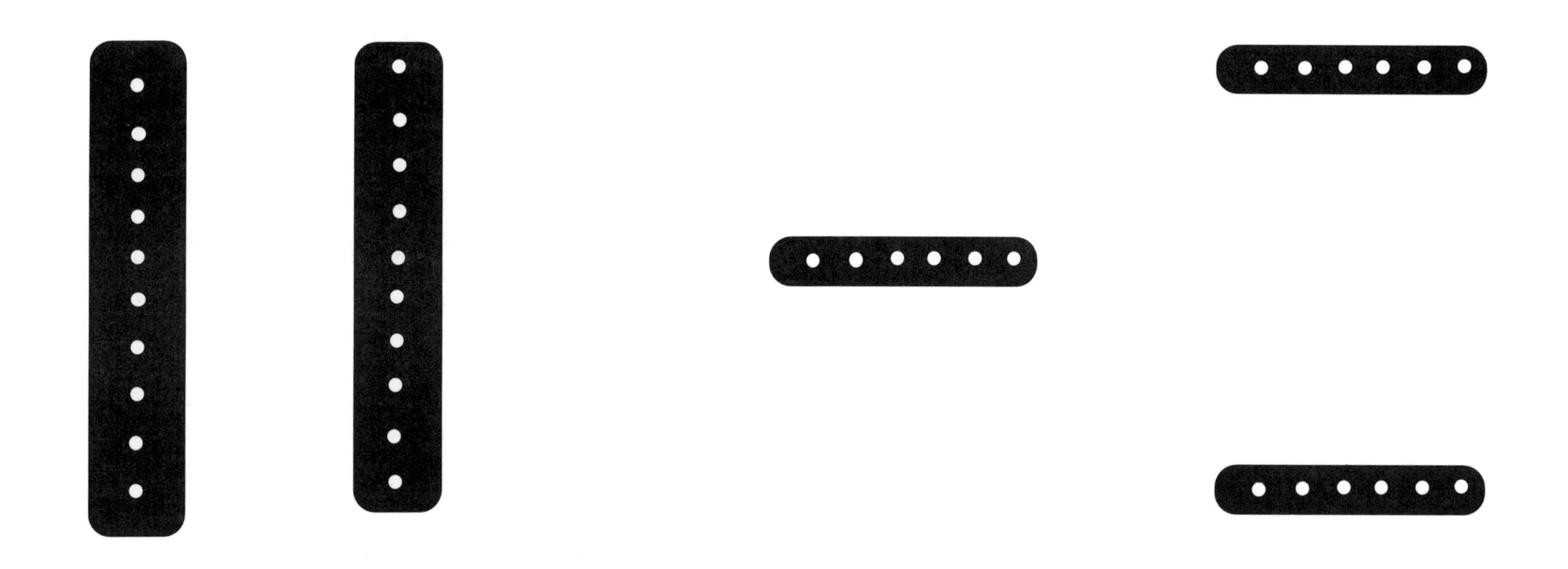

George Platts

Finally, after we attended five years of engineering school all on a co-op basis, we graduated in 1931. We found ourselves in the Great Depression. Jobs were scarce. However, a recruiter from the General Electric (GE) Company came to interview prospects. He offered jobs to three of us—George Pettibone, Richard Steves, and myself. We were flattered.

George Pettibone accepted and spent his entire career with GE. Richard Steves declined because he wanted to stay in Cincinnati. I said I would accept if I could get a job in radio. The recruiter said that all new electrical engineering recruits went into the "test course." This course involved checking out large motors and generators before shipping them to the customers. However, he agreed to talk to a "Mr. Wilson" who was in charge of radio to see if he would take on a new electrical engineer.

I suppose the depression made GE's radio business poor, hence Mr. Wilson had no opening for me. Perhaps some readers will remember that when radio first went into production, there was an arrangement between GE and Westinghouse to make products marketed under the name of RCA. Eventually, the federal anti-trust division insisted that this was a monopoly, so the GE and Westinghouse facilities had to be turned over to RCA, which became an independent company.

So I job hunted with the idea that I would take almost anything remotely connected with electrical (preferably radio) engineering until I could get what I had always wanted. I started off working with a one-man firm that sold high quality electric motors and mercury switches.

This hardly related to radio. but it at least brought in a pay check. After about six months, I found out that a man who had run a retail radio and phonograph business (which went bust in the depression) was going to represent a company which made sound systems. This seemed a bit closer to what I wanted, so I went to work for him.

Unfortunately, he was still broke. That meant long hours and little pay. But, fortunately, I had a neighbor who worked for the Crosley Corporation. It had started out as a mass producer of radios, then added ranges, refrigerators and stokers. My neighbor got a job for me there.

I worked on the development of auto radios and a power supply that could convert the six volt battery supply in a car to the high voltage for the plate circuits of the vacuum tubes. But the best part was not only did the Crosley Corporation make all this stuff; it owned a 50 KW radio station as well.

Powell Crosley was not satisfied with 50 KW, he wanted something much bigger. So, he applied for and got a 500 KW experimental license for WLW, then bought a RCA transmitter (at this stage still being made by GE and Westinghouse). This had a high power stage to attach to his existing 50 KW Western Electric transmitter. The high power stage consisted of a radio frequency power amplifier receiving its audio signal from a power modulator.

No sooner was this put on the air than Powell Crosley applied to operate a 500 KW full time. The Federal Radio Commission issued the full time 500 KW license—and WLW became *The Nation's Station*.

The Chief Engineer decided that the signal did not have the quality the *Nation's Station* should have. We immediately began work on improving the modulator. This was a task involving going to Mason, Ohio. That's where the transmitter and its eight hundred and thirty-one foot antenna were located. We made all possible preparations in advance of twelve midnight. As soon as WLW went off the air, we began our work. We had only a few hours because everything had to be put back together by six a.m. when WLW went on the air. After many nights of this short schedule, we had the quality our Chief Engineer, R. J. Rockwell, deemed acceptable.

John Robert Smith

Four years later in the throes of the depression (1931), I received my BSEE degree from Purdue. I was not among the less than ten percent of my class graduating with an engineering job in hand. The next four years were spent working any kind of job available. During that time, I borrowed seventy-five dollars to get married, became a Fuller Brush salesman, a schoolteacher, and a GMAC accountant in a Chevrolet dealership. By 1936, there was some easing of the depression.

That was the year I obtained my first engineering job. That summer, Oklahoma Gas & Electric Co. (OG&E) gave me a temporary job surveying people living in rural areas to see who would apply for electric service if .it were available. The survey was to prove or disprove the feasibility of rural electrification. At summer's end, I was cleaning the 120-volt main substation control battery, then calibrating new watt-hour meters, and later residential meters in the surrounding small towns. A year later I was sent to a division town where I designed 4.16 and 13.8 KV distribution system extensions to serve new outlying customers.

Paul D. Andrews

I graduated from Penn State with a BS in electrical engineering. It was expected that I would return to work for Ma Bell, but when I found that meant working in the Bell Telephone Laboratory in New York City, I rebelled. Having worked two summers in New York City, I wanted nothing to do with it. My credentials were good, and I had five firm job offers.

I finally wound up with General Electric in Schenectady, New York. Of course, they said I had to spend a year or two on their "Test Course," which I refused to do. I told them I had no interest in turbines, motors, and such. I wanted to go directly into radio engineering. The recruiter said that the Radio Department was only a little operation. That it would never amount to anything (he did not know Dr. Walter R. G. Baker). Finally, I got just what I asked for.

I reported for work at Schenectady in June, and was assigned to the Low Power Transmitter Section, of which Irving F. Byrnes was the Section Leader (now called Manager). I received the magnificent salary of twenty-seven dollars per week, which was considered good. The "Test Course" engineers only received twenty-four dollars per week. We all worked a forty-four and a half hour week.

John L. Shaw

After demobilization, I did a year of graduate study on a Wisconsin Alumni Research

Foundation Fellowship in Physics and operated their mass spectrometer. At an American Physical Society meeting in January, 1947, I was offered a job with the International Nickel Company (INCO) in their Research and Development Applied Physics Section. My specialties became the use of nickel in vacuum tube cathodes, nickel containing magnetic alloys and constant modulus alloys. The IRE show was our big exhibition each year where I manned the INCO booth—Dad was at RCA's. My ultimate compliment came the year he jokingly complained he was getting to be known as John Shaw's father (Dr. G. R. Shaw, IEEE Fellow) instead of vice-versa.

Charles T. Morrow

After getting my doctorate (shortly after WWII), I accepted a position at the Sperry Gyroscope Company on Long Island. During my stay, Sperry achieved what may have been an all-time low in personnel relations for all companies. It had been a specialty company. The war had forced Sperry into activities that the management did not understand or trust—electronics, systems, and manufacturing. By the end of the war, however, management perceived that there was money in these activities.

The production plant became the focus of the company's operations. All activities, including research, were carried out according to production ground rules. The company was reorganized—the marine effort, which had been the mainstay of the company, became a small department lost in a ponderous organization. Experts that had been recruited during the war hastened to leave. All gyroscope development was officially stopped except for that of the tuning-fork gyroscope, which I was soon to direct.

This development, spearheaded by Joseph (Joe) Lyman, was sold to Sperry as competition to the gyroscope program of Professor Stark Draper at MIT, then the country's primary source for inertial guidance. Joe claimed that Sperry's new instrument would be free of the errors that plagued the wheel gyroscope—static forces would have no effect.

Joe had arrived at a configuration that for reasons he did not understand had much potential, but each new instrument constructed drifted badly from the configuration. It became accepted that there must be some one source of drift. Each member of the development group worked independently, coming up with successive designs according to his own whim, with erratic uncontrolled changes of detail. Each dreamed of finding the one defect that blocked the way to glorious success.

Joe told me he wanted me to ease myself into the leadership of the group. Once I had proved myself, my responsibility would be made official. He did not tell me that Roland Barnaby had been and was still the leader. He evidently also did not tell Roland what he had asked me to do. I soon found myself in a vicious fight that was not Roland's fault. With a Navy development contract approaching, the fight had to come to an end and some order had to be put into the efforts of the group.

Finally, I told Joe that if he wanted me to lead the group, he would have to make it official and take his chances. He acted. He told me I could fire Roland if I wanted to. I chose not to do this, but I took tight control of the group. I began an organized program of identification and elimination of possible errors in the instrument. Understandably, all the engineers of the group resented this—each had lost his personal dream. Roland was shattered, but in a few months he returned to productive participation. He had many essential talents. In the next several years, we brought the tuning-fork gyroscope to the edge of inertial guidance performance.

By 1951, after five years on Long Island, traffic congestion, especially on summer weekends, had become intolerable. Furthermore, the Sperry engineers had formed a union, largely because of the working conditions. But these conditions did not turn out to be negotiable. Consequently, the union negotiated for money instead. The more the union succeeded in this arena, the worse working conditions became. Finally, I accepted a job with the Hughes Aircraft Company (which did not make aircraft) in California.

One morning I wandered into Joe's office wondering how to tell him I was leaving. He looked up at me and said, "I'm leaving." I said, "So am I." We had a hilarious day. The Navy arrived in the afternoon to discuss possible follow-up contracts. No one in the company was in a position to make a commitment.

Emil Gaynor

I was offered a scholarship by Westinghouse for a master's degree. However, I decided it was not fair to my wife so I signed up for night school and went to work. During the three years it took me, I worked as a research and development engineer for the ADT Company and the Fairchild Guided Missiles Division.

There it was my very good fortune to work on ferrite microwave devices, and the first transistor circuit to fly on a U.S. missile (the Lark). Because transistors were so new, we had to do our own testing and devised a test set that was the first on the market. Fairchild was not really interested in making them. So we built and sold twenty-five ourselves and then dropped it.

With some prompting by a colleague who had gone to RCA, I applied and got a job at the Moorestown facility. There I took the lead in building a hybrid fire control computer for the Talos weapon system. It was made of relays, crossbar switches, and the usual complex of servos and resolvers. After that I was sent to the West Coast facility, now General Dynamics, to work on the launch control system for the Atlas missile. My next assignment was back in Moorestown to design and build the various display systems for the Ballistic Missile Early Warning System (BMEWS). I became a third level manager.

We built the first computer to operate with no scheduled downtime. It was a very successful design thanks to some very talented designers and enough government funding to allow adequate testing of the components. As the program neared completion, I was assigned to complete a large space chamber at RCA, East Windsor—that was exciting. Looking for new programs became a major part of my work. This was my first exposure to marketing, which would evolve into my later career activities.

Frank T. Luff

Jobs for graduating engineers were very scarce in 1948 and 1949. I took the *only* offer I received. It was from the Rural Electric Administration in Washington, DC. This was a far cry from my radio and radar experience but I was married, with a family, and needed a job.

Four years later my job as a power engineer was abolished and I was assigned to the new Rural Telephone Program. This was a blessing in disguise because my job title was changed to "electronic" engineer which allowed me to pursue other jobs in the electronic engineer category.

Four years later, in 1956, I applied for a job as a radar engineer with the Civil Aeronautics Administration. Because of my previous interest and experience in radar, I got the job and was sent to the CAA RADAR School in Oklahoma City, Oklahoma. I did very well and, a year later, was promoted to Supervisory Electronic Engineer, GS-13. That was pretty good in those days because the highest GS grade was GS-15.

Harold W. Lord

My decision to apply for the job of test engineer at the General Electric Company and my acceptance, I believe, was influenced by Dr. Sorensen. On September 13, 1926, I left for Schenectady, New York, by ship from Pasadena, California, going through the Panama Canal with a half-day in Havana, Cuba. I arrived in time to start my job on October 4, 1926.

My first "test" assignment involved the so-called "neutralizing" of the two radio-amplifier stages so that they did not oscillate. My second job, starting after the first of January, was in the Research Laboratory Section which was to develop new vacuum tubes for radio receivers.

I was offered a permanent job in that section, which I agreed to take after completing a "test" assignment in inducting motors and automatic switchboards. On October 10, 1927, I began my job as an engineer of the General Electric Company.

My first assignment was to establish, in a house on the outskirts of the city, a test station for evaluating new and developmental radio tubes in RCA and competitors' receivers. At that time, RCA was owned jointly by the General Electric and Westinghouse (sixty percent by GE and forty percent by Westinghouse). RCA was formed as a Patent Pool, thus avoiding a lot of litigation and need for extensive cross-licensing agreements.

The location of the test station was well away from manufacturing plants and so fairly free of electrical noises in the power supply and through the air. I would frequently sit up late at night, after the 50KW local station, WGY, went off the air. Then I could pick up broadcast stations from as far away as the West Coast.

The first type of mercury-vapor thyratron became available about this time. Also, the General Electric hermetically-sealed-motor-and-pump electric refrigerator became available. This refrigerator could not be used in those areas of New York City supplied with only 125 volts direct-current (dc). My section manager, Mr. W. C. White, thought that thyratron inverters might be developed to operate the GE. refrigerator from 125 volts dc.

To demonstrate this possibility, he asked me to design and have the shop build a small inverter. I designed an inverter to deliver enough power (about 40 watts) to operate an eight-inch electric fan. After studying some technical papers on the subject, I designed a parallel-type inverter, including the transformer and inductor, and had it built. The inverter did a fine job of operating the fan from a 125 volt dc source, in this case a small motor-generator set. This experience sparked my interest in industrial electronics.

Leo Berberich

In 1931, the Great Depression was rapidly deepening. The news was that Ph.D.s were accepting farm jobs and I, growing up on a farm, knew all about farm jobs. I did have the dubious distinction of being offered a position by a General Electric Company interviewer at their Pittsfield, Massachusetts, High Voltage Laboratory. However, I was telegraphically fired within three days due to the rapidly worsening business conditions.

Fortunately, Dean Whitehead came to my rescue and saved me from the pitchfork and the manure spreader by offering a research assistant's position at Johns Hopkins University. I was to work on the Dean's major project, the study of the life of high voltage oil-impregnated paper insulated cables. Considering the times, my salary of one hundred twenty-five dollars per month was almost princely for a poor recent graduate. And it "kept the wolf from the door" until something better appeared on the horizon.

After about four months, something better did arise in the form of a letter from the Director of the Mobil Oil Research and Development Laboratories in New Jersey. Mobil requested that Dean Whitehead recommend one of his recent graduates for a new position they had created. Already the leading supplier of transformer oil for the electrical industry, Mobil wanted to expand its insulating oil business by developing the special oils required to produce high voltage lead-covered oil-impregnated paper cables and oil-impregnated paper power capacitors. I "just happened to be available" and gratefully accepted the position at two hundred and forty dollars per month. With one assistant, I proceeded to set up the laboratory to carry out this project.

So that I might better understand the language used by the some eight hundred chemists and chemical engineers in the laboratory, I decided to take a night school course in organic chemistry. The background gained from the course helped enormously, as did the extensive literature of the American Chemical Society.

My stay at the Mobil Research and Development Labs lasted about six years. During that time, I set up and organized the laboratory facilities for my project and developed several promising petroleum oil candidates for high voltage cables and power capacitors. Pilot quantities of these oils were tested by both General Electric and Westinghouse. However, they did not make a contribution to Mobil's bottom line because the Monsanto Chemical Company had just announced the polychlorobiphenyl synthetic liquids known as "PCB." PCBs soon replaced all petroleum products in power capacitors and even partially replaced petroleum in some transformers. These liquids were used extensively until, as we all now know, they were found to be carcinogenic and environmentally dangerous.

K. L. Rao

I had wanted to be a physicist. Soon after graduation at Madras, I went to Cambridge (United Kingdom) for higher studies in physics. I took Honors in 1935. I soon realized that job prospects and research facilities in our country were meager. But then the broadcasting service in India started and there was a great future.

I went to the Imperial College of Science and Technology (City and Guilds College), London for study in advanced telecommunication with emphasis on broadcasting and related matters. I returned to my country in 1939 and, after knocking on several doors, I joined as a technical assistant (perhaps the lowest rung in the ladder of technical men) at the All India Radio in 1941.

The All India Radio Service expanded when the Second World War broke out. A Planning and Development Unit was formed to work out the details. I was chosen as one of the engineers to assist in working out the expansion plans.

The story may, perhaps, not be complete without touching upon some items of work which I was called upon to execute apart from my main duties. These extra duties came up during WWII when equipment and components were in short supply. I became, as it were, a problem solver.

In the 40s, India had yet to build a radio industry. Some equipment was received from abroad. Radio receivers in large quantity were received. They were sent by sea and were badly damaged in the process. These receivers were serviced and brought back to working condition.

During the installation of the broadcasting house at Delhi in the 40s, a similar situation arose. Studio equipment, viz., pre-amplifiers, line amplifiers, etc., were imported. These amplifiers had a poor S/N ratio and fell far short of broadcast requirements. These amplifiers were over-hauled and had their wiring re-arranged to bring their performance up to required broadcasting standards.

Hyman Olken

This company's main product was a machine that measured the thickness of a rubber sheet while the sheet was still in the pasty, uncured state. At that time, inner tubes for automobile tires were a major rubber industry product. They were produced by rolling out a wide sheet of unvulcanized rubber, cutting a narrow strip off that sheet, then putting its two long edges together to form a long cylindrical tube. You then butted the two ends of that cylinder together to form a circular tube, the initial form of any inner tube. Then this tube was vulcanized by heating and you had a completed automobile tire inner tube.

If the raw rubber sheet was rolled out to the right thickness, the inner tube lasted a year and the tubes sold well. If rolled out too thin, the tubes lasted only a few months, or even only a few weeks, and the customers never bought that brand again. If the sheet was rolled out too thick, the resulting inner tube lasted for years and very few were sold.

To measure the thickness, the sheet was passed through a two-piece capacitor connected in a 100 KC resonant circuit. The circuit current was measured by an RF (Radio Frequency) current meter. The meter was calibrated for the current corresponding to a sheet of the right thickness.

If the current was too high or too low, the

too-thick or too-thin raw rubber sheet was not used to produce inner tubes. To provide control of the sheet's thickness, the final drum over which the sheet passed in the production process was heated by steam entering to the center of the drum. The steam heated the drum and its expansion produced a thinner sheet. So all you had to do to get a sheet of the right thickness was to turn the valve that admitted steam to the last cylinder of the machine that rolled out the raw rubber sheet.

Of course, this could be done by an automatic control for the valve which admitted the steams to the drum, but an easier way was found. We hooked up an RF milliameter that measured the current in the tuned circuit. It was a strip chart recording milliameter. The chart was marked with two lines along the length of the chart, to indicate the top and bottom limits of the thickness of the rubber sheet.

An operator was stationed to operate the steam valve so the thickness of the rubber sheet remained between these two lines. He was paid on a yardage basis for all the sheet that fell within the two lines. For any sheet of thickness that fell outside those two lines, the operator was not paid. The resulting control of the thickness of the rubber sheet was perfect.

This job lasted for nearly two years when the depression started to catch up with the inner tube industry. Some people had to be laid off. They decided to lay off the single people first, so I was caught in the layoffs

Franklin Offner

In 1934, I went to the physics department of the University of Chicago. I started work in microwaves building magnetrons. After a few months, I found the professor in charge to be completely incompetent, and was going to shift to work with Professor Robert Mulliken in quantum mechanics. However, I ran short of money. When I was offered a job with Shure Brothers Company as an engineer to work on microphones I accepted.

I started at twenty dollars a week (fifty hours), because the president of the company said he "didn't know if I was capable of handling the job." After a month, the chief engineer, Ralph Glover, said I was obviously better qualified to be chief engineer than he was, and that I should take over his job and he'd shift to sales. I told the president that since I was obviously capable of handling the job, I wanted a ten dollar a week raise. He refused and I immediately quit. (This was the only job I ever held in industry.)

This turned out to be the best thing that ever had happened to me. I had been advising one of my radio amateur friends at the university (where I continued to live) on how to build an amplifier for biopotentials for the laboratory of neurophysiology. My friend was trying to help Professor Gerard, a prominent professor of neurophysiology. When I returned to the university that evening, my friend told me they had expected me to quit and that there was a job waiting for me as a research associate under Professor Gerard.

I soon got the amplifier working and for the first time saw nerve action potentials. No one understood their origin at that time, so the field immediately attracted me. I resolved to get my doctorate in biophysics and find out how nerves work. (A problem on which I would work, with some interludes, for the next fifty years!)

Electroencephalography (EEG: brain potentials picked up between two points on the scalp) was just starting to be studied at this time (end of 1935). We started to record the EEG of humans by recording the trace of a rather faint CRO (cathode-ray oscillograph) on moving photographic paper. Thus, we couldn't really see what was going on until the paper was developed—often several days later. I solved this problem by inventing the direct writing oscillograph, which recorded potentials up to and over 100 hz with a pen writing on a moving paper tape.

Another problem was that amplifiers, at that time, all had "single ended" input potentials being referred to a common reference. Therefore the EEG could not be recorded between independent pairs of electrodes, which we wished to do—for example, to compare potentials on the two sides of the head. I solved this problem by inventing the differential amplifier, with common mode negative feedback to insure true differential amplification.

Together these two inventions permitted the simultaneous recording of a number of separate potentials. This allowed the experimenter to follow them "on-line." A great help to the worker. But the differential amplifier had another unforeseen advantage: it drastically reduced electrical interference from power cords and such. This eliminated the need for placing the subject in a shielded room, as had previously been required.

Thus, the first clinically useful EEG was born. But at that time, the full clinical value of the EEG was not yet recognized, while the value of the electrocardiograph (EKG) was well known. I was thus advised to adapt these two inventions to the EKG. I proceeded to develop a portable direct writing clinical EKG, the forerunner of the present EKG apparatus.

Robert Newman

Very few firms were recruiting on campus in 1936, yet I felt secure: I had a job with Industrial Rayon Corporation in Cleveland. What's more, I had taken several courses in textiles and rayon manufacturing. Then the axe fell across my neck: Industrial Rayon went bankrupt.

In late August, 1936, while visiting the Cleveland College employment office, the clerk asked me if I could fill a request for Glass Technologist at General Electric (GE). I said, "Of course."

She then called someone at GE and made an appointment with a Mr. Wilhelm Kahlson* for the next Monday—this was Thursday. I rushed to Case Western Reserve to see if they had a glass technologist on their staff. "No, but Professor Winkelman at Western Reserve was well known in the field." I found him in his office. He agreed to take me on as a *special* student and spend two hours with me on Friday and all of Saturday morning. I enrolled in the graduate school, bought the four textbooks Winkelman suggested, and read them in time for my Monday appointment.

In spite of not having more than four hours sleep total for the previous four days, I got the job. Twelve hundred dollars a year for a five and a half day week, with Christmas, New Years Day, Thanksgiving and Independence Day as paid holidays.

One of my early assignments was to find a

better method of "guaranteeing" a specified bulb quality level. For example, bulbs required uniform side walls or they might be too weak, or not seal properly to other pieces of glass in the automated equipment.

It seemed to me that much of the bulb problem was due to the non-uniform travel of the glass through the four hundred foot long glass furnace. I sold Bill Kahlson on a project which included building a transparent Lucite model of the furnace, so we could "see" what was happening to the glass flow.

Soon I found out that I had "bitten-off" more than I had expected. Scale-modeling turned out to be a very complex job, and had not been attempted before. For example, to obtain the conditions I wished to observe, mathematical models soon convinced me that some of the model's dimensions required the cube root of the scale and some the square. And some of the temperature differences required the square root of the scale.

I built the furnace using viscous corn syrup dyed to absorb the radiant heat from the electric heating used to simulate the gas flame. The model was unstable—as was the furnace it was modeling—and led to improved dimensioning of the next furnace, greatly increasing the stability of the glass flow. Mind you, this novel research, with little assurance of success, was done during a period when profits were almost nonexistent. Kahlson was pleased with the results and made me Works Engineer at a salary of twenty-four hundred dollars a year.

The next fun project began while I was taking a group from the neighborhood boys club through the Telling-Bell-Vernon Dairy's pasteurizing and bottling plant. They had just installed a large colloid mill to make homogenized milk. I thought that perhaps such a machine might permit cheaper materials to be used for etching—or "frosting" lamp bulbs, while giving more uniform results. It worked, reducing materials cost by half and improving appearance and strength.

At about this time, I began to become aware of the "we versus they" syndrome. "We" being the Lamp Department and "they," the rest of the company. "We" earned money, "they" spent it. "We" couldn't make customer brand lamps because "they" wouldn't let us. "We" sold fewer lamps if "they" manufactured low quality toasters. General Electric was not one happy family.

**Wilhelm "Bill" Kahlson, a naturalized Finn, had jumped ship in 1929 and used his engineering background to get a job at GE. When he started rising in management, and GE found that he was an illegal alien, a way was found to have Congress pass a special bill to make Bill a citizen. At my interview, he was manager of the Pitney Glass Works. He died about 1970.*

Thomas A. Nelson

After receiving my BS degree in electrical engineering, I went back to work for the Los Angeles Department of Water and Power. I did not attend the campus interviews by General Electric (GE), Westinghouse, etcetera, because I wanted to remain in the Los Angeles area. So what happened? I stayed for three years working in overhead distribution design. There I designed facilities for the huge post-war housing tracts in the San Fernando Valley and the industrial area around the Los Angeles International Airport.

Meanwhile, I took evening classes at the University of Southern California and obtained a master's degree in electrical engineering. Then I found myself working for DWP at the Westinghouse Steam Turbine Works near Philadelphia, Pennsylvania, GE at Schenectady, New York, and other eastern U.S. factories, as well as in Europe, as a resident inspection engineer.

Kenneth Miller

My first job, and I mean job, at twenty-five cents per hour, was to wire floor lamps in a dingy loft operation in Chicago. This was quickly followed by becoming a phonograph needle inspector looking for pits on the iridium tips welded to the long life needles produced by the Permopoint Company. This was not for me.

Good fortune was at my side and I was hired by Zenith Radio at 6001 Dickens Avenue in Chicago in 1940 as a production trouble shooter. This time at forty-five cents per hour. You learned fast....with the requirement that you troubleshoot defective all-band home radios at the rate of eight per hour! Among the products we produced was the famous "Radio Nurse," a children's nursery monitoring system that caught the public's imagination following the kidnapping of Charles Lindbergh's son from his nursery at his home.

Then the call of the West was heard loud and clear. So I hopped into my 1939 Chrysler. (Talk about diesels, this horseless carriage burned one quart every hundred miles! No kidding.) Eighteen hundred miles and twenty quarts of oil and fifty-nine hours later, I arrived in Los Angeles. Driving down Pico Boulevard, I paused at 11919 Pico Boulevard as I noticed an interesting antenna atop this old one story stucco building. A ham radio operator for eight years, I had developed an affinity for roof top antennas that appeared to be "senders" as well as "receivers" of RF (radio frequency) energy.

This building turned out to be a former dairy complete with milking stalls still intact in the rear of the structure. With little effort you could almost catch the aroma and hear the sounds of its former occupants, the milk cows. Only now it was the West Coast headquarters of Lear Incorporated. I entered and there was Bill Lear cussing because the Lord had only given him two hands and this was at least one hand shy of what he needed at that moment. He was trying to repair a LearRecorder.

This was the famous wire recorder produced by Bill Lear's company following the acquisition by our armed forces of wire recorder technology developed during World War II by the Germans. Lear produced portable and large models for the home market. It turned out that the unit Bill Lear was working on (and did not have enough hands to repair) was a unit that the Los Angeles police department had purchased for telephone monitoring purposes.

Not knowing I was just a casual visitor, he turned to me for help, within ten minutes he hired me. Thus I began a wonderful friendship and a multi-year career with Lear Inc. Also, again, later when I rejoined him as Vice President and a member of the founding management team at LearJet in Wichita, Kansas.

At Lear, I was responsible for the development of the first automatic pilot sold to the general aviation market—the famous Model L2. Later, amongst other accomplishments at

LearJet, I had the opportunity to introduce and manufacture the world's first mass produced eight track tape players. We produced many thousands of these players, and hundreds of thousands of the eight track tape cartridges for both home and automotive entertainment at a plant we started from scratch in Detroit.

These were really fun years....and as Bill Lear and I said to each other so many times, "I feel guilty for being paid to have such a great time"...but never so guilty that we refused our pay checks!!

Vernon McFarlin

June 15, 1931—graduation into twenty-five percent unemployment. If one had a college degree and a dime, one could ride on the Boston Elevated Railway.

I had worked part-time jobs and borrowed money to train myself for skilled employment. After four years of struggle, graduating into a world where jobs were practically non-existent was an embittering experience. I will never know what it feels like to be offered even one job upon graduation from college. Although I was unemployed for a total of only about five weeks during the Great Depression, the work I had paid only subsistence wages.

Simpson Linke

During the three years of my graduate study, I was a member of the "service course" staff and taught basic electricity, machine theory and laboratory, and basic electronic circuits to non-electrical engineering students. I completed my graduate work in June, 1949 and spent the summer months of that year at Brookhaven National Laboratory.

I worked with Professor Henry Hansteen on research and development of a three-phase linear induction motor that was used to pump mercury in a closed system. We were able to achieve an overall pumping efficiency of one percent. The laboratory conducted a patent search and discovered that in the mid-twenties Albert Einstein and Leo Szilard had been granted an all-inclusive patent for electrically driven mercury pumps. Since their efficiency was only one quarter of one percent, we congratulated ourselves on a job well-done. Later we learned that an efficiency of 14 percent had been achieved at Argonne National Laboratory with an induction-motor pump that used liquid sodium as the linear "rotor."

After that exciting summer, I returned to the campus as an assistant professor. I taught machinery and electronic circuits to senior chemical engineering students for the next two years.

Tom Cartin

I graduated in 1950 at the height of the Korean war. I was in the Army Reserve so I went to work in the research and development laboratory at Fort Belvoir, Virginia. I hoped that I could make a deal to stay there if I got reactivated. Things cooled down so I went to work for the Glen T. Martin Company.

I joined a new group there working on a new thing called a digital computer. The design was based on flip-flop tube circuits. With brilliant foresight, I didn't see much future in that field so I switched to the electrical design section. It's amazing to realize how simple decisions, easily made at the time, profoundly affect one's life.

Yardley Beers

While I was ending my graduate school work, there was a great shortage of jobs for physicists. I must have made at least fifty applications, sometimes paying my own expenses for job interviews. Just as I finished my experimental work, positions opened up because of the imminence of the WWII.

New York University needed an emergency replacement for someone who had left for war work. I got the position partly because I was readily available and partly because Martin Whittaker (later director of the Clinton Laboratory in Oak Ridge) had known me from various meetings. This job was better than many of those for which I had been rejected! I spent the year 1940-1941 at New York University learning how to teach and writing up my thesis.

Paul Burk

I was stationed at Langley Field when WWII ended. Upon discharge, I was hired by the NACA Instrument Research Division. My first assignment was to miniaturize the existing 100 kc oscillators using proximity fuse tubes in the existing Franklin oscillator circuit, (a two stage resistance coupled L-C oscillator). After running some tests on this circuit, it became obvious that it was very unstable due to the "Miller Effect" which I announced as a passing remark to one of the other engineers. Well, all pandemonium broke out. What was "Miller Effect"? A mad scramble to some reference books followed.

When I assembled and calibrated this RM 1 telemeter, I had no idea what we would be discovering in that forthcoming flight at Wallops Island, off the coast of Virginia. The RM 1 was powered by two tandem cordite rockets and had fixed aileron settings to cause the missile to roll counter clockwise during its flight. It was tracked by 584 radar, fastax camera and four channels of telemetering, azimuth (roll), total head pressure, lateral and transverse acceleration.

The first record to be developed was the telemeter recording from the mirror galvanometers. As you might well imagine, all hell broke loose when our roll record indicated that the missile, which was rolling counter clockwise at takeoff up to about mach 8, reversed its roll and was rolling clockwise during transonic flight.

"Telemetering must have screwed up again, etc. etc." is all we heard until some time later when the fastax film came out of developing. It showed that the missile did in fact experience a reverse controllability. Consternation throughout resulted, especially from 'ole Marve Pitkin who was the project engineer on the RM 1.

About a week later, the analytical group came up with the same result (but I still wonder if it was independent of our experimental data). Flash! At precisely this time, Chuck Yeager was at Edwards Air Force Base ready to take off in the Bell. A panic call went out to Edwards to call off the flight until the NACA group analyzed the data and determined a proper flight procedure for Chuck. Needless to say, the esteem of our telemetering group went up a few notches after this flight. Later that year, NACA telemetering was written up in the Congressional Record, (1946-47).

Leslie Balter

I wound up in a civil service position in the Signal Corps General Development Laboratory at Fort Monmouth, New Jersey. We were getting ready for World War II and, at least, the Signal Corps was hiring engineers indiscriminately.

I was looking forward to research and development in the Signal Corps. I soon learned that the military industrial complex we are concerned about today, did in fact, have its roots in those days. The lab was little more than a paper agency which supervised contracts to industry for the actual research and development. The lab's function was primarily testing, supervision of some production testing, and supervising contracts placed with industries such as Bell Labs, Western Electric, and Ferris Instruments.

I was assigned to the Vehicular Installation Division of the Mobile Radio Section. The section chief was a civil service career draftsman who had absolutely no concept of radio communications. He also had no interest in the young college graduates coming into his department. We were concerned with the installation of mobile communications in command vehicles, trucks and tanks, including half tracks and full tanks. My particular job involved what is now known as RFI (radio frequency interference) suppression. In those days, it was taking the radio noises out of the vehicles.

Herbert Butler

Reluctantly, I started to look for a better paying position and a more promising future. I answered an ad by Wallace & Tiernen, a company that designed and manufactured precision meteorological devices and also very profitable beer dispensing equipment!

Of greater interest, the company had a contract with the Signal Corps Engineering Laboratories at Fort Monmouth, New Jersey, to build a substantial number of balloon-borne radiosondes. They were seeking what today would be called a quality control engineer. I had a very good interview with the chief engineer. I was offered the position, but I had had enough of inspection and testing. I wanted to be involved much more in the research and development of new devices, and this would not be possible. With some regret, I did not accept the offer.

Some time later (June, 1941) out of the clear blue sky, the mail brought a letter from the Signal Corps Engineering Labs, urging me to arrange an interview for one or several job opportunities in research and development at the Labs. I should have realized that this was not a fortuitous coincidence. However, it was not until later that I learned that Wallace & Tiernan had recommended me to Fort Monmouth. I started with the Signal Corps Labs in August of 1941, working on developing radio direction finders to track balloon-borne radiosondes.

John E. Duhl

The Navy Air Development Center recruited me at college. I really really enjoyed my job Although, as my first assignment, getting an entire radar system to check out for compliance with specification nearly overwhelmed me. I coped with the assignments as they came and left in late '53 because my mother was failing, and I thought it better to be at home. Westinghouse, who had previous contact with me at the college interviews, provided the opportunity.

My assignments at Westinghouse were equally steep. I was given a test bench and some magnetos to experiment with for a week, maybe two, when the supervisor came by and asked if I could handle the test. Thinking, he meant the bench, I said I could manage it. "Fine," he said, "I'm moving Al, the test engineer, out so the department is yours—ten benches—three shifts and about forty opera-tors." A foreman was there but the engineering decisions were mine.

About a month or so later when I felt in control, he came back and said, "After Monday, the exhaust department is yours, too." With the price of a tube at over two thousand dollars back then, I figured I would do what I thought best. If I was wrong, it wouldn't go unnoticed for very long.

Joseph E. Guidry

June 7, 1933, I got my master's degree in electrical engineering from Tulane University. With my coveted BE degree in those bleak days and being the only recipient of an engineering offer from the General Electric (GE) Company, Schnectady, New York, I was quite elated and optimistic. This euphoria, however, was short lived because GE withdrew the offer because of the Depression. Interim employment was obtained with the Shell Petroleum Corporation, and the R.P. Farnsworth Construction Company.

Lester E. Haining

Born on a farm in southwest Nebraska in 1913, I had never met an electrical engineer or seen a factory or a place where such people worked until I started college. However, I liked to experiment with electrical things which I read about in an encyclopedia at high school and in radio magazines. I remember thinking that it must be great to work at some place like General Electric or Westinghouse all day and get paid for doing it! I finally started college at the University of Nebraska at age of twenty-five. (Drought and the depression made it impossible before then.)

The timing turned out to be fortunate. I graduated from college during WWII with a BS in electrical engineering. So in the Army, I was assigned to the Electronics Training Group, was sent to the Radar School at Harvard and MIT—an excellent school experience—and wound up with a Signal Radar Maintenance Unit in the southwest Pacific.

C. Richard Ellis

After high school, I attended a radio trade school and obtained a First Class Radiotelephone license. With this, I was hired as a technician in a broadcast station. I thought I knew a great deal, and when the station decided to install a remote studio in another town, I agreed to build the phone line amplifier for the studio.

The distance to the remote was about fifteen miles, so I decided that a pair of 6L6 tubes and an output transformer with a 600 ohm output, providing about 30 watts output, would be about right. It was about right to put every telephone in La Grange, Georgia, out of business due to crosstalk.

The phone company took a dim view of this technique for increasing the station's coverage. The problem was solved with a VERY large T pad. I continued in the broadcast business (by not repeating the same mistake) until the U.S. entry into WWII.

My first year as an engineer was spent mostly in the factory, learning the real world. It was probably the most valuable experience of my career in electronics. In the factory, the efforts of engineering, drafting procurement, production, and inspection converged on the process of "getting it out the door." The strengths and weaknesses of a large organization, and the vital role individual human beings play in the process, becomes apparent to the new engineer. I learned that an elegant mathematical solution might be very difficult to machine and assemble.

Robert S. Dahlberg

When I graduated in 1931 receiving the BA degree with High Honors in physics with a math minor, I wanted to go into research. I had my eye on Bell Labs. But Bell wasn't hiring and neither was anyone else. I continued in school at the University of Texas.

In that distant day and time, it cost little, if any more, to go to school than to sit around the house. I was earning most of my expenses as a student assistant in the physics department. There I earned a magnificent fifty dollars per month grading papers and lecturing first year lab sections. Even the state of Texas was broke in that depressed era and I was paid in warrants which the state called in whenever it collected enough money.

I made good progress toward my Master of Arts degree. Course work was finished and my thesis well under way in June of 1932 when one weekend a rumor came that the Humble Oil and Refining Company was hiring. Believe me, Monday morning I found myself sitting on the door step waiting for them to open up! They took me on and my career was launched.

My working years began nursing the first crude reflection seismograph amplifiers in the semi-tropical undergrowth of the Texas/Louisiana Gulf coast. Today geophysicists measure the motion of continents in centimeters per year.

William A. Edson

After a long hiatus, Bell Labs resumed hiring in 1937. Partly, on the basis of my brother's talents, I received a job offer and joined the Systems Development Department in August, at the princely salary of forty-two dollars per week. No time clock and only thirty-five hours a week! During that summer I toyed with an offer from the fledgling General Communication Company, but wisely chose Bell Labs. It may be worth noting that the words "electronics," "communications systems," and "development" were not yet in the vocabulary of the general public.

Elias Weinberger

When I graduated from high school, it was January, 1937 and I had just turned sixteen. The depression was still around in all its glory and things at my house were pretty bad economically. I wanted to go to college and my parents wanted me to go. I registered at

Brooklyn College. I had made up my mind that I was going to major in chemistry and that is what I did. However, when it came to buying the necessary books and paying the necessary lab fees, I found the money required was not available.

It almost broke my parents' hearts but I managed to get a job in a factory earning about ten dollars per week and transferred to night classes. During the next two years, I earned approximately twenty-three credits. Working in a factory for eight to ten hours a day and trying to stay awake in lectures was just too much, I finally had to call it quits. I was earning about twenty dollars per week by then and was the main breadwinner of the household.

The next big event was WWII. I had just turned twenty-one and had applied for a job in Washington, DC. In January, 1942, I was called to Washington where I started to work for the War Production Board as a clerk. I was drafted into the Army in December, 1942. After going to school for electronics (where I also did some teaching in math), I was sent overseas in March, 1944 to join the 51st Fighter Control Squadron. Working with "Merrill's Marauders," we chased the Japanese out of India and then out of Burma when the war ended in August, 1945.

During the time spent in the China, Burma, India (CBI) Theater of operations, I wrote to a wonderful young lady for about two years. Upon arriving home in 1946, I took a trip up to New York where she was living at the time. We met for the first time on February 14, 1946, St.Valentine's Day. On June 23, 1946, we were married and went to Washington, DC, to live.

My first inclination was to forget about my promise to myself about finishing college, but my wife wouldn't hear of it. After being out of school for about nine years, I registered at George Washington University, College of Engineering, under the GI Bill. To my surprise, I was accepted in February, 1947. My original plan of being a chemical engineer had been modified; I was now going for electrical engineering in electronics. Graduation came in May, 1950 and I got my BEE degree which required one hundred and fifty-five credits.

Throughout my Army career (and now in my university work), I kept specializing in one aspect of electronics, which was communications. That was how I wanted to apply myself. I began looking for a position in the field of communications electronics, but 1950 seemed to be a slump year for hiring electronic engineers. I tried all of the big firms but all I was offered was a job as a lab technician. I kept turning these down. After traveling up and down the east coast, my wife and I returned to Washington. I was hired in September, 1950 by the Civil Aeronautics Administration as a communications engineer. I was put in charge of the "Overseas Foreign Aeronautical Communications Stations (OFACS)."

Fred J. Tischer

As often happens in academia, the first assignment of my career was that of a teaching assistant while toiling toward my Ph.D. at the University of Prague in Czechoslovakia. It was a shocker.

My task was preparing the experiments in electricity and magnetism shown by our much admired electrical engineering department head during his class. After a few unsuccessful tryouts and letter writing to my predecessor, I realized that deceptive tricks had to be applied to get most of the experiments going. The tricks made them work not the phenomena explained so elaborately and convincingly by the professor in class. This experience was really depressing and disappointing. However, it let me concentrate my job hunting in research and development. I completely disregarded teaching as a career. Research and development thus became the mode of my professional endeavors during the following years.

Olaf B. Vikoren

At the end of 1924 with my diploma in hand, I returned to Norway hoping to land a good job there. I soon learned that the economical conditions were so bad that no new generating stations were being contemplated for a long time. The disappointment was nerve-racking.

Also, there were practically no available jobs of any kind, anywhere in the country. Since foreigners seeking employment in Germany were not welcome, there was no sense returning there.

The most likely possibility for success was to try one's luck in the United States. I arrived in New York on February 16, 1926. Among other companies, The Philadelphia Electric Company (as it was then called), was advertising for technical help. It was recommended to me that I should try to get a job with that organization. I followed that advice and was hired as a draftsman on March 1, 1926.

The job was far from being glamorous, but I knew that an immigrant ought not be choosy in his first job and experience was what I needed. After having been a draftsman for about a year, I was transferred to Field Engineering and sometime later to a inside job in the Engineering Department.

Edgar W. Van Winkle

There were no openings when I graduated in June of 1936, but through Rutgers University placement, I was offered a position at the Best Manufacturing Company in Irvington, New Jersey, designing loud speakers. I started at fifty cents per hour. I was raised to twenty-seven dollars and eighty cents per week when they found I was making forty to fifty dollars per week at the fifty cents per hour rate.

While there I gained a knowledge of loud speakers, production test equipment and assembly line production procedure. I enjoyed the work there, I ended up as a production supervisor for over two hundred people. We went through a vicious strike and it was exciting times but I felt it wasn't engineering.

Then I was offered a position as electrical engineer designing and testing electronic equipment at Western Electric (WE) in 1937. However, the depression of 1938 caused WE to terminate all engineers who had started in 1937 or later. This was two weeks before I was due to be married and I had no advance notice.

I was fortunate to find placement as an electrical engineer in the Department of Labor of New Jersey. A thorough knowledge of electrical problems in industry and the experience of working with hundreds of different types of industry was gained here. The opportunity to compare them was afforded me while I was obtaining my Master of Science degree in industrial engineering with a minor in electronics.

Shortly after I received my degree in June of 1943, I was offered a position at the Allen B. DuMont Lab., Inc., at 100 Main Ave., Passaic, New Jersey. I was hired in 1943 to organize and set up a Standard's Laboratory.

Marvin Udevitz

Unfortunately, 1949 was a recession year and my joy in being a college graduate was short lived upon finding there were few jobs available for electronic engineers. Since I couldn't find a job in what I considered my speciality, I reckoned that anything of a professional and technical nature was suitable. This was particularly true since I had used up every cent I had, along with the so-called "GI Bill," to get through school.

And so it was in the summer of 1949, I accepted my first engineering job with the United States Bureau of Reclamation in the Field Construction Division of the North Platte River District with headquarters in Casper, Wyoming. At first blush, one might consider this situation as incongruous. Why would an electrical engineer with an electronics major get involved in field construction even if it were "electrical" in nature? Certainly, the work was more civil engineering oriented than electrical—no?

Indeed, it was fertile ground for a civil engineer, but it turned out that taking that job with the Bureau was one of the very best decisions I ever made. Two reasons why: the first, although it didn't become apparent to me until many years later, everything that one learns is important and the time will surely come when each piece of knowledge and every skill acquired serves a need.

The requirements of the job were a far cry from what college had prepared me for. The experience I gained was probably the very best experience in maturity that a new grad could have. It provided me with a piece of fatherly advice that I could comfortably pass to youngsters fresh from school who ended up in organizations which later I managed.

The second reason was probably more self intertwined. I do believe that every engineer worth his or her "slipstick" (I guess the right word today would be "lap top computer") wants to do something which will leave his or her "mark" on the world. You know, something which is markedly different because "I was there." A lot of folks spend their whole careers hopefully looking for that circumstance and it never happens. But in my case, it not only happened, but it happened during my first job experience out of college.

I had moved from the field construction area into the District Office reporting to the office engineer when he was offered the position of field engineer on the Eklutna Project. This was the first major hydroelectric development in Alaska. When he went I went.

Thus, in the fall of 1950, I found myself driving up the Alcan in anticipation of the big adventure. When I got there I found that I was among the very first on-site and I was initially assigned as the Chief of Surveys, again hardly a job for an electrical engineer. But what fun! Later on I was assigned chores more in keeping with my background, i.e. building the switchyards, substations, powerlines, and so forth.

In the time I was there, I was able to lead my crew in establishing the tunnel axis as well as the portal locations, setting up the tunnel in-line surveys, and establishing the location of the substations in Anchorage and Palmer in addition to supervising their construction and activation along with the 110KV interconnecting powerline—Alaska's first. Eventually, many folks contributed to completing the project and bringing everything on line. But only a few had the privilege of being there when there was "nothing" and leaving when there was "something." And I was one of them!

Theodore Schroeder

In my senior year, I could opt for either an electrical enginering or a civil engineering degree. After much weighing of the pro's and con's electrical engineering won out.

The wisdom of this decision was highlighted in my mind when—on February 19, 1936—one of the foremost life's occurrences took place: I was recruited and hired by the General Electric Company to come to Schenectady, New York, and join the student engineering training program, commonly known as the "Test Course." The remainder of my senior year, in that post-depression but still job-scarce era, was one of high elation.

My first assignment was in testing thyratron control drives for gun directors for U.S. Navy battleships. This struck me as odd for someone interested in power utilities, but it turned out to have an advantage: I learned more about electronics than I did when in school.

J. Rennie Whitehead

My initial area of specialization was determined by the exigencies of war. I was to graduate in 1939. Before that happened, I was approached in the spring of 1939 by my professor, P.M.S. Blackett. He told me that he was on a scientific advisory committee to the British government and that a world war was imminent. He strongly advised me to let my name be put forward for a job in a secret research establishment once I had graduated.

I agreed and, later that summer, found myself being interviewed in London by Robert Watson-Watt and Charles Percival Snow. I didn't know their histories at the time, but Watson-Watt was, of course, credited with the invention of radar in the mid-30s. C.P. Snow, who was then head of the Civil Service Commission, became famous for his series of novels, *Strangers and Brothers*.

So in August, 1939, after graduation, I joined the Air Ministry Research Establishment at Bawdsey, which was effectively the birthplace of radar. I must have been about the twentieth member of the professional staff. By the time I joined, two-thirds of the coast of Britain was already protected by radar and airborne radar was undergoing its first flight trails.

On the first of September, 1939, fearing that Bawdsey would come under immediate attack from the air—conveniently situated for the enemy as it was on the southeast coast of England—we moved lock, stock and barrel to Dundee. A couple of moves later, we became TRE Malvern, located in the Malvern Boys' College, in Great Malvern, Worcestershire. By then we were about twenty-five hundred strong.

TRE was a marvelous research organization in every sense. As a result of the shortage of electrical and electronic specialists, it had recruited top scientists and engineers, almost independent of discipline, who could bring fresh minds to this new technique of generating and manipulating short pulses of energy. We were all very young and the sky was the limit in taking responsibility. We also worked very hard because we knew in our bones that radar was essential to survival, as indeed it proved to be. Therefore, we gained a great deal of technical and managerial experience in a very short time.

Early in the war, I was given the responsibility for the design of the IFF (Identity Friend or Foe) Transponder that went into all allied ships and aircraft and, later, responsibility for the entire separate-band radar identification system (IFF Mark III). After the Tizard mission revealed the secrets of TRE to the United States government, a small U.S. contingent of senior scientists came to live in Malvern for a year, before returning to the U.S. to the MIT Radiation Laboratory. They included Isidore Rabi (who was awarded a Nobel Prize in 1944). Years later, when he and I were on the NATO Science Committee together, we would exchange reminiscences of Malvern.

We worked very closely with the Royal Air Force in the development of radar equipment for their use. We had our own airfield with our own fleet of military aircraft of every type, flown by service officers but maintained by civilian crews. I did a lot of flying in the early years of the war testing the results of our laboratory efforts.

H. R. Weiss

Upon graduation from Lowell Institute School, I was given a job at MIT's High Voltage Research Laboratory. The lab did research in million-volt Van De Graff generators and the X-ray tubes to go with them. The research was mainly for the treatment of malignancies with, what was at that time, the best radiation equipment for that purpose. I was a technician, later on promoted to D.I.C. (Division of Industrial Collaboration) staff member. The highlight of my experience was helping design a two million volt generator which I eventually helped install in Philadelphia 's Oncologic Hospital together with my friend, Charlie Goldie.

While I was happy with my work and liked the collegiate atmosphere of the laboratory, I soon began to realize my technical shortcomings as I compared my knowledge with that of Charlie, an MS from MIT or Bob Cloud, another MS or Dr. Trump, a D.SC. The result of my feeling inferior caused me to seek employment with Raytheon, a company made up of engineers, radio amateurs, and former members of the Submarine Signal Corps.

I believe now that I fitted right in with a group of technical people, many of whom lacked a formal engineering education, but I didn't know it then. For instance, my good friend Dennis Picard started as a technician about the same time as I did as an engineer: Dennis's education had been RCA's TV service school (and he was an excellent TV repairman). He was promoted to be an engineer without more technical education than I soon after I had left the company. Seven years later, he became a manager over the design group that we had belonged to. Eventually, Dennis was promoted to be one of the directors of Raytheon; whether his education had been augmented, I don't know.

My feelings of inferiority continued while working at Raytheon. During an interview with General Electric from Utica, New York, I was promised consideration to study towards a master's degree at Syracuse University if I accepted employment with them. I must state that while working at MIT and Raytheon I had taken many graduate courses at night at Northeastern University (NU) with excellent grades. However, I was told by NU that under no circumstances would they consider granting me any degree regardless of my excellent record. Now, here was Syracuse University willing to give me credit for all these courses and grant me a Master's degree to boot if I took a minimum number of courses.

I started work in Utica and received a Master's degree in Engineering in three years, a result of part time study which General Electric was instrumental in facilitating: they gave me time off to study. The thesis subject was one of my projects at work and they even helped me by typing my thesis.

G. B. Welbourne

After four years at the University of Illinois over a six year period, my first employment as a graduate engineer was as a construction electrician foreman. In 1950, the best paying electrical engineering job offered through the university placement was about forty percent of what was available to me from a former employer as an electrical construction foreman.

My employer was a large national engineering and construction company. They soon added as much responsibility to my job as was possible for me to handle. Gradually the job required more technical ability and less practical experience. Interestingly, though, the salary structure was still based on practical experience and ability to perform construction activities. Home office graduate engineers made less than good concrete pouring foremen.

E. A. Weiss

My first job, after graduating from Lehigh in 1940, was with the RCA Manufacturing Company in Camden, New Jersey. RCA called its newly hired electrical engineering graduates "student engineers" and put them through a two year program of floating from one department to another, two months in each department. The program gave the young men (there were no women and an equal number of blacks), an opportunity to see twelve different departments. It gave the departments an opportunity to see a number of new hires.

RCA did not expect to keep us all but probably hoped that those who left, like those who left the General Electric "Test Course," would find it easy to spell "RCA" when writing purchase orders for later employers. I suspect that student engineers were carried on the budget of the training program itself so the served departments got free but highly educated lab assistants.

The first thing I learned was that the company was not devoted to full disclosure as far as new employees were concerned. Although living costs did not surprise me, young engineers from the South had not been told that our salaries (all BS's got one hundred twenty dollars a month, all MS's got one hundred fifty dollars a month) would not go nearly as far in Camden, New Jerey as they would in Georgia. Jobs were hard to get so nobody left right away.

Price Wickersham

My first job in 1943 was in the experimental development department of a major aircraft engine manufacturer. There I was involved with engine test instrumentation at one hundred seventy bucks a month. I remember the janitor wouldn't take me up on my offer to trade paychecks.

But my first in-depth job experience was after I got out of the service and joined Midwest Research Institute in Kansas City, Missouri. As a central technical facility for a wide variety of midwestern industries, we were challenged from all sides to solve an incredible variety of problems. Here's a short list of only some of the projects I had during the ten years I was there:

- An instrument to measure uteral contractions and cervical distensibility in the birth process.
- In 1950, we attempted to develop a hearing aid imbedded in eye glasses using some of the very first Bell Labs point contact transistors. (They came in cute little felt-lined boxes.) This project failed. The transistors were too damn noisy.
- An airborne target drone to measure the angle and miss distance of .50 caliber machine gun bullets.
- A servomechanical weighing device for high speed filling of coffee cans and cottage cheese containers in a production environment.
- An early microwave (1948) in a sandwich vending machine, using a WWII radar magnetron. This project flopped because juices in the meat would make the bread soggy.

It's a little funny, but some of the project failures were as much fun as those that succeeded.

William H. J. Kitchen

Christmas Day, 1930 at the age of eighteen, I embarked on a Leyland Line freighter out of Liverpool. I was bound for Bermuda to take up the post as Father's engineering assistant at the Bermuda Railway.

The "clutch" gave more trouble than any other part of the motor coaches, so I worked out an improved system utilizing a Westinghouse Signal solenoid connected to the air brake receiver and the clutch cylinder. This eliminated the "clutch control valve." Only two handles were now required to control the gear box, the selector and the throttle.

I was so sure that it was a good idea that I approached Father. (I was always required to think out exactly what was to be said before approaching the him with an idea, and to think out the consequent answer should it not be relevant.) The answer this time was, "If it was a workable idea, someone other than you would have used it already." At once I decided to wait until Father went to the United Kingdom on leave, make the contraption and fit it. If it worked as well as I expected, I would at least get a raise in pay. It worked, but instead, I was unceremoniously fired. I left the Railway the same day after five years service.

Although I visited home every Sunday for dinner, the Railway was never discussed. As time went on, I heard that the valve (which my father had "blown off") of my system was fitted onto each coach until all the original rolling stock had been modified. It was greater success than even I had imagined. Some time later, a letter from the Chief Engineer, my father, to Westinghouse Brake and Signal Company commented on the solution to the clutch problem that had been found. My system was then outlined.

In 1946, the Plymouth Locomotive Works produced a sixty-five ton diesel locomotive, using epicyclic gears and an electro-pneumatic control system along the same principal as the Bermuda Railway modification. The "Flexomotor" unit was described in the handbook *Diesel-Electric Locomotive*, written by the editors of *Diesel Power and Diesel Transportation*. I have been informed that they are still available.

Edmond S. Klotz

Housing was almost impossible to find on Long Island, but I heard through the grapevine that there was a new development going up, later known as Levittown. They were offering veterans the first opportunity to rent or to purchase. I opted to rent a brand new house complete with range, refrigerator, and an automatic washing machine for fifty-five dollars per month!

At Sperry, I worked in the Armament Radar Department on the antenna and microwave portions of a long range tracking radar which used monopulse techniques and a metal plate lens antenna. It operated at C-band and there was very little test equipment available at those frequencies. I ordered test equipment from the Polytechnic Research and Development Company because I could get much better delivery than I could from our own Sperry Instruments Division. Needless to say, management took a very dim view of this but I prevailed.

I shared a cubicle of an office with Lou Cutrona, who later became well-known for his work on synthetic apertures. Among other talents, he was an excellent chess player and I never beat him during our lunch hour matches. He had a remarkable memory. When our match was unfinished at the end of lunch hour, we would put the board away and Lou would set the pieces back up entirely from memory the next day.

Working for a large company had many advantages but there also were disadvantages or, at least, minor inconveniences. I had a grievance filed against me by the union that represented the electricians because I replaced a blown fuse myself. Much of my antenna work was done on the roof and I had to call the professional riggers to move my equipment from the laboratory to the roof. There were two groups of riggers, however, vertical and horizontal, and coordinating their efforts was time consuming.

After several years of interesting work, I became aware of the "salary curve" which plotted years of experience vs. salary. I realized that I would never become wealthy at Sperry and began to look elsewhere.

Thomas M. Austin

I graduated from Colorado University in 1932 at the bottom of the depression. None of the graduates that year received a job offer. I was fortunate enough to received a graduate assistantship at Iowa State College, where I obtained my MSEE degree in 1933.

From 1933 to 1936, I had five short jobs with little engineering content, interspersed with periods of unemployment. In 1936, I finally obtained a job with the Southern Colorado Power Company in Pueblo, Colorado, starting as a "grunt" on a line maintenance crew, at forty cents per hour. From there I progressed to power plant switchboard operator, then to draftsman in the vice president's office.

In 1939, the Roosevelt administration relaxed the requirement for a political endorsement for all jobs and offered a civil service examination for engineers. I took this examination for engineers and obtained a job with the U.S. Bureau of Reclamation in October, 1939, as a junior engineer.

George H. Barnes

At Harvard, I had no conscious vocational goal. I took math, physics, and circuit design courses because they seemed important. My master's degree was a way-station in a doctor's program which I didn't get to finish, which is probably just as well. I was headed toward a thesis on antenna theory, so I might have been stuck with solving Maxwell's equations forever.

For my first job, the Franklin Institute hired me as a physicist to hold electromagnetic opinions about the reflections of radar pulses from statistically described terrain. On the same contract, they needed to build an analog recorder of radar returns, and a correlator and spectrum analyzer to process them. They had no engineers available so they picked me instead for on-the-job training.

I learned to use a soldering iron. I learned logic design by reading Kiester, Ritchie and Washburn. I learned circuit design tricks from MIT's *RadLab Series*. I learned spectral analysis, including covariance spectra, stationary times series, and how to use operational amplifiers from co-workers. I've been an engineer ever since.

M. Lloyd Bond

The first job I landed was as a test engineer (sic) at Cornell Dubilier—testing X-ray system capacitors. (This was in 1938.) The training was minimal—enough to accent the safety provisions of working with tens and hundreds of thousands of volts. This was emphatically NOT what I had expected—but I had to eat! On eighteen dollars a week, I paid room and board, came home on weekends, and even had enough money (on occasion) for a date.

However, a few months on the job, with complete boredom arising, led me to quit. For a little while, until I could land a better job, I fixed radios for a living. Then, in 1939, I became an associate in a consulting firm that designed broadcast antenna systems and performed proof-of-performance tests for broadcast stations. Now, this was *real* engineering. My boss was attentive and caring, teaching me the nuances and the shortcuts, the necessities for doing a good job. The pay was GREAT (forty dollars a week)!

W. A. Dickinson

My college mentor, Dr. Percy H. Carr, referred me to his friend and former student, Dr. Robert M. Bowie, who was director of a laboratory developing television picture tubes, at Hygrade-Sylvania Corporation in St. Marys, Pennsylvania. This lab reported to the Receiving Tube Engineering Department at Emporium, Pennsylvania. It was set up at the request of Philco Corporation, Sylvania's major radio tube customer, and worked closely with Philco Engineering. At the time, televison seemed "just around the corner."

Dr. Bowie made me an offer to start as a junior engineer, in July, 1937, at sixty cents per hour. I accepted at once and spent the next forty-two years in the Sylvania's organization, working in design and development of cathode-ray tubes. (Hygrade-Sylvania became Sylvania Electric Products, Inc. and was later bought by GTE.)

But television didn't get started commercially as promptly as we expected. England, which had taken the lead in development and in broadcasting, got involved in World War II and shut down television activity. United States radio manufacturers turned their attention to FM radio, which promised quicker return on their investments. So, early in 1940, our laboratory closed. I felt fortunate to be retained on the payroll and was transferred to the Receiving Tube Engineering Department in Emporium, Pennsylvania.

Paul G. Cushman

After graduation from MIT and Middlebury in 1941, I joined the General Electric (GE) Company and stayed with that company for forty-two years and five months. In those days, new engineers at GE entered the "Test Course" where they gained practical experience testing GE products in several different departments. After that, those so inclined could enter, in addition to the Advanced Engineering Study Courses, the Advanced Engineering Program for another one to three years for a series of engineering assignments in various departments.

However, World War II was on and the engineering rotation assignments were forced to stop. The young engineers were "frozen" into military departments. I was assigned to a small group that was designing and manufacturing selsyns and other small electrical equipment. Even though small and low cost, these devices were crucial components in many of the gun directors and gun drives and mechanical computers that GE was supplying to the government.

This assignment was very interesting and instructive. I remember one rather minor event with particular clarity because it was the source of a moment of great technical satisfaction but with an ending of disappointment. GE unions were on strike and we non-union engineers could not get through the picket lines to work. We were getting paid and were told to "think about useful projects" at home.

I spent part of my time analyzing the torque response of selsyn differential motors, which, for reasons not understood, had considerably less torque than expected. The analysis showed that the leakage reactance of the rotor should be increased. This counter-intuitive result was not believed by my boss or my associates. But since it wouldn't cost much to try it, my boss gave the go-ahead.

So when the strike was over, I ordered a differential selsyn with a rotor ground undersize but with a thin layer of soft iron wire wound round the rotor to give the desired leakage reactance. Tests revealed an increase in torque of the expected amount.

So I submitted a patent docket, fully expecting that the company would certainly patent such a novel idea. But it didn't. Selsyn motors of any sort were not being designed into systems because new accuracy requirements dictated that selsyns be used in their electrical signal mode, with follow-up servos providing the needed mechanical torque. Thus, my patent idea had no future business value, which was much more important to the patent filing decision makers than the novelty of the idea.

Hans K. Jenny

As I was growing up in Switzerland, especially during in high school, I started thinking seriously about what profession to chose. I was already in a high school branch leading to a technical education, (as concentrated to liberal arts, business, medicine, etc.). In Switzerland, the university for the best engineering education was the Swiss Federal Institute of Technology, the MIT of Switzerland, which incidentally had close ties to MIT.

So I started studying electrical engineering at the Swiss Federal Institute of Technology in 1939. Since military service is compulsory in Switzerland, I also started basic training just about the time the Germans marched into Poland and launched WWII.

At graduation time, I received my MS in electrical engineering (dipl. el. ing. ETH) with the highest ranking in the class and on the exams. I was invited to join Professor Tanks' staff as an assistant professor and to work towards my Ph.D. (I also received offers from two other professors.) For the latter, I was assigned the field of microwave tubes and soon became engrossed in velocity modulated tubes with which I had become acquainted through an article in the *Saturday Evening News* by the Varian brothers. It did not take me long to have the first klystron operating in Switzerland.

At this point, WWII was nearing its end and having been closed off in our small country by the surrounding German armies for four years, a strong desire had developed to see the world. I especially wanted to visit the United States, where so many electronics advances were taking place. I applied for a visa at the American consulate and in December, 1945 was told I had a three month window to emigrate to the U.S.

I abandoned my Ph.D. efforts and found myself on a thirty-two day journey to the U.S. aboard a liberty ship (freighter). Twenty-four hours after I landed, I was offered a job at the RCA Electron Tube Division in Lancaster, Pennsylvania, a company with which I subsequently spent my entire career!

Chapter five

Catching a break

Alfred J. Siegmeth

In 1940, I joined the Jungarian Philips Corporation, a subsidiary of the Dutch Philips Company. I acted there, on a corporate level, as the engineering director of two manufacturing plants. In addition, I formed an engineering research team and coordinated the development of a demonstration unit of a small radar system for the Hungarian military.

Toward the end of World War II, in November of 1944, the German military dismantled our radio transmitter tube factory and transferred the hardware and the associated employees, including our family with two small children, to Vienna, Austria. As the war came to an end, we moved our team in a western direction in three consecutive steps from Vienna to Munich, Germany. Fortunately, we avoided the military zones occupied by the Russians.

The cannibalistic dismantling of our transmitter tube production facility by the German military and our forced transfer with our employees propelled us to the lowest point in our lives. The positive side of this action was the fact that it saved our family and our employees from a meaningless life under a communistic rule. To improve our family's food supply, for almost two years, I repaired many inoperable radio sets in the farmlands of Bavaria and bartered my electronic knowledge for food staples.

In 1949, the U.S. Congress passed an Immigration Act for displaced persons. After we received an invitation from our U.S. sponsor, Mr. Dave Marx, president of a radio company in Albany, New York, we knew our dreams and plans to enter the United States would be realized. Despite my early difficulties with the spoken English language, in 1949, I started my new engineering career in the U.S. as the Chief Engineer of the W.T. La Rose Company, in Troy, New York.

I developed there, during four years, a line of specialized, dielectric heater models destined for the plastic and rubber industries. They were used to enhance manufacturing productivity, and the quality of plastic and rubber products.

During 1953, we moved to California and I returned to my field of telecommunications. In Los Angeles, I joined the Resdel Engineering Corporation. (In 1954, we received our U.S. Citizenships with dignity and honor.) At Resdel, I coordinated the system engineering efforts for the development and production of precision, passive and active Doppler type, launch vehicle tracking systems, called DOVAP and UDOP, for NASA and the U.S. Air Force. My appointments were Chief Engineer and Senior Vice President.

We handled only government contracts; thus, consequently our work load was never stable enough to keep our key employees. During my thirteen years there, we sailed through hectic and unpleasant firing and hiring phases. I have found that most of the other small electronics companies had the same experience.

In 1966, the Jet Propulsion Laboratory (JPL) invited me to act as the Deep Space Network Manager of the Pioneer 6 through 11 missions. In 1969, with the cooperation of a newly formed Pioneer Support Team, we initiated, developed and implemented for the Deep Space Network (DSN) all additional capabilities specified in our system engineering plan for the successful tracking and data acquisition support of the Pioneer 10 and 11 spacecraft. They were mankind's "first" fly by visitors of Jupitor and Saturn, the largest planets of the solar system.

Nick Petrou

Returning to the States in 1946 after the war, I had a choice of joining the Westinghouse Electric Corporation, where I had been an engineer in training, or RCA. I settled on Westinghouse because the company offered ten cents an hour more than RCA.

In 1951, I moved my family to Baltimore, Maryland, where Westinghouse assigned me to start up the Defense Electronic operations. In 1961, I was sent to the Advanced Management Program (AMP '39) of the Harvard Business School. I became General Manager of the Aerospace Division of Westinghouse in 1966. I advanced to President of the Westinghouse Defense Division and later to Executive Vice President of the Westinghouse Defense Space Group. In 1977, I was Corporate Vice President of Human Resources at Westinghouse headquarters.

John J. Di Nucci

The U.S. Department of Energy was created during the Carter administration on October 1, 1977, at which time WAPA (Western Area Power Administration) also came into existence for a five year period from 1979 to 1983. I was assigned to serve as Chief of the Division of Power Contracts and Rates in the Office of Power Marketing Coordination under the Assistant Secretary of Energy. One of my duties was to assist and coordinate the efforts of the rate staffs of the five power marketing administrations in setting in place several power rate increases to assure timely repayment of the various federal hydroelectric projects. My job was to assure that the rate orders were properly prepared for approval by the Secretary of Energy and the Federal Power Commission.

The most difficult part of this assignment was to make the upper echelon of the Department understand the legal responsibilities of the five power marketing administrations in the sale of wholesale power to the thousands of municipal, state and cooperative electric utilities. (The five entities are the Alaska Power Administration, The Bonneville Power Administration, The Southwestern Administration, and the Western Area Power Administration.)

There are several acts of Congress governing the preference in the sale of federal power and the rates which can be charged. These laws must be strictly followed. This whole process took several months, so the first rate increase was not put into effect until the middle of 1980. Once, the Department heads achieved that understanding, however, things proceeded very smoothly.

Upon completion of the FPC's National Power Survey at the end of 1963, I was faced with the choice of returning to New York or remaining in Washington, DC, in some other capacity. I chose to accept a position with the Bureau of Reclamation, and as a result of adding several new hydroelectric projects and the expansion of several existing ones, my career began to escalate. Although I felt at first that there might be some risk involved, I soon realized it was the best decision I had ever made.

In looking back over the years, I find that there were several accomplishments which could be considered noteworthy. The two that readily come to mind, however, are the development of a timely procedure for the adjustment of wholesale power for federal hydroelectric projects, and the complete repayment of several of those projects during my tenure. The former was accomplished with considerable help from the office of the Solicitor of the Department of Energy.

Perhaps my worst failure occurred while with the FPC and was not to be realized until a decade later. That was my inability to predict the drastic reduction in the rate of growth of the use of electric energy in the U.S. which resulted from the "oil crunch" caused by the adverse OPEC actions in 1973. It would have been much better if I had used the method employed by the Edison Electric Institute by considering an ideal scenario and a worst scenario. This would have given a wide range of possibilities, with proper qualifications for each. Ah, if foresight were as accurate as hindsight.

John L. Shaw

My second career at INCO (the International Nickel Company) was a total switch. INCO had been working on airborne prospecting methods. I, being the only electronics engineer in the company (we had several power electrical engineers), was sent to Toronto for two weeks to work with consultants to try to solve an impasse they had run into. It turned out I had an idea that worked. The two weeks stretched into eleven years.

We were the first to develop a successful airborne electromagnetic (EM) system. It was flying for five years before a competitor had a working unit. In the early years, I was the operator until I trained a team to take over. As a result, I have flown at five hundred feet over most Canadian Provinces and Territories, Minnesota and the outback of Australia. Our biggest discovery was the Thompson, Manitoba ore body. This I believe is still the world's most valuable single ore body. Engineering turned out to be a very exciting field.

The success of the Airborne EM system gave me a reputation for organizing new departments. When INCO wanted to make the Huntington (West Virginia) Rolling Mills independent of the New York office, I was transferred to set up a technical service department to aid customers in the use of nickel alloys; thus, began my third career at INCO. My responsibilities expanded to include establishing a market development group to find new applications for our existing alloys and a market research group to find what alloys would be needed in the future. The department was up and running nicely when I got a call to come see Harry Wingate, Chairman of INCO in New York.

Harry wanted me to "investigate the possibility of mining nickel from the bottom of the ocean"—another major career change. It was undoubtedly the greatest project an engineer could have. I learned about the field by a couple of crash courses, a lot of library time, and visits to any company that could help us (one hundred and sixty thousand air miles that year). I presented a report to the Board that the subject appeared worth pursuing.

I got the appropriation I had requested and an okay to move to the west coast as our work would be in the Pacific. By then I had a staff of four: an oceanographer, a naval architect, a mechanical engineer and an administrator. We all moved to Bellevue, Washington. in the fall of 1971. We first leased an exploration vessel and later outfitted our own to survey the bottom of the ocean, eight hundred miles south of Hawaii, where it is three miles deep. We also designed an ocean mining system and a refinery to support it while doing economic studies.

Everything looked good except the price. It was going to cost over fifty million dollars a pilot test to prove the system could work. At this point we recognized the need to broaden our international representation. Besides the cost and high technical risk, the UN (United Nations) was showing an interest in the "riches of the sea." Dean Ramstad, an INCO Vice President, and I spent 1974 on airplanes putting together a joint venture of our Canadian INCO, Ltd., with German, Japanese and U.S. companies who could contribute technology, trained people and money.

In May of 1975, we signed an agreement with three German companies, three Japanese companies (with twenty additional Japanese companies as honorary participants), and a U.S. oil drilling company to form Ocean Management Incorporated (OMI). I was made President and General Manager. The whole project is worthy of a book, which I will write someday, but for this paper suffice it to say that the cooperation among our companies was absolutely amazing. Each sent their most capable people with the talents we needed and gave them total support. We expanded our operation in Bellevue. In three years, we designed and built a system that was the first to successfully mine nickel containing nodules from three miles deep. In a half an afternoon, we picked up eight hundred tons!

However, sadly, the UN was circulating a treaty, "The Law of the Sea," to establish a company named "The Enterprise" which would be given all developed technology (including ours) at no cost to them. The company would pay no taxes even for their refinery and their employees would pay no taxes to any country. For us to obtain a claim, we would have to prove out twice the area needed and give the Enterprise the data for them to choose which half they wanted.

Can you believe President Carter was ready to sign it! You can imagine how that went over with the Boards of the joint venture companies—our prudent course was to document what we had learned and shut down the operation.

I was made an INCO Vice President and we moved back to New York. My new responsibility was to plan all the rolling mills and forging shops. I was also on the Planning Committee for the parent company, INCO, Ltd., where my experience in prospecting and mining and contacts in Japan and Germany were valuable. I was also Chairman of the small venture companies set up to develop laboratory ideas into nickel markets.

We carried out a substantial reorganization of how the rolling/forging mills were managed. When that was done another move was suggested. My wife of forty-five years and I had moved twelve times and we decided that was enough so I took early retirement. We moved back to Bellevue, Washington, since it was our favorite of all the places where we have lived.

I had been retired less than two weeks when a group of my ex-ocean mining colleagues asked me to manage a small company they had formed when we shut down OMI four years before. They were building a side scan sonar that was second to none in performance. For me, it was quite a change from being an INCO Vice President—with lawyers, accountants and researchers at my beck and call—to handling all the business details, with the help of one executive assistant. It was a hard sell since each sonar cost as much as a million dollars. Also, our potential customers were nervous about our small size. However, we managed to do okay until we landed a two million dollar contract. To finance it, we ultimately had to sell out to a large company.

Homer M. Sarasohn

I was a twenty-nine year old radar expert at Massachusetts Institute of Technology when I received a telegram from Washington in 1946 summoning me to Tokyo. The telegram said I was wanted by General Douglas MacArthur.

I thought it was a joke. However, I soon found myself enroute to Japan.

MacArthur, in charge of the U.S. military forces occupying the defeated nation, wanted to restart the Japanese radio industry as soon as possible. This was so the military could communicate with the dispirited population.

Japan was in rubble, as were the companies that built electronic gear. Workers had scattered and their machinery had been carted away to the countryside.

Since I once had worked for a U.S. radio manufacturer, I was assigned to the Civil Communications Sector, in charge of all forms of communication for the occupation forces.

My first task was to send teams to find the electronics industry's production equipment and round up workers. Then I had shacks built for makeshift factories. We, literally, ran the operation, the industry such as it was.

Immediately there were problems. The first batch of vacuum tubes, turned out by the primitive plants, had a ninety-nine percent defective rate.

To the Japanese running the factories that was not unusual. The nation's industry managed to supply its war machine through sheer volume of production. They did not understand the idea of quality.

I called in the plant managers and said this was intolerable. I then asked the managers to identify one problem they could work on to improve quality.

There was utter silence. They were not expected to make meaningful contributions to their companies in this sense.

Then the managers began talking among themselves. My interpreter told me they were deciding what answer to give that would most please the American.

At that moment, I decided to study Japanese language and culture to help break down this communication barrier. I also decided to become a sterner teacher.

In a certain sense, I became a dictator. It was necessary to do. The Japanese, used to taking orders from a militaristic government, were good followers.

Within nine months, the plants began producing radios, and output later picked up. But quality remained inferior. In 1949, a colleague decided the Japanese plant managers needed a course in American management.

But some members of the occupation forces opposed spreading U.S. production know-how to the Japanese. "We would create a monster," they warned.

I argued that Japan must be set on a strong economic foundation or it would become a long-term drain on U.S. taxpayers. Finally, it went up to MacArthur. At a meeting, the opposition, from the occupation force's Economics and Social Section and I, each gave a twenty minute presentation to the General on our viewpoints.

MacArthur got up and began to walk out of the room. As he neared the door, he turned around and told me, "Go do it." That was all. Those were his orders.

Soon my colleague, Western Electric engineer Charles Protzman and I, spent a month assembling a textbook. Then we taught a course to the Japanese plant managers.

The course stressed that quality must be a commitment of the entire company, from the management down, and must take precedence over profit. The course also taught that the managers must listen to their workers and gain their trust. In addition, the managers were told that every company needs to compile a clear statement of its purpose, a company motto of sorts, that sums up its reason for existing.

Among my pupils were Sony co-founders Akio Morita and Masaru Ibuka, Matsushita Electric's Masaharu Matsushita, and Mitsubishi Electric's Takeo Kato. As you know from history, they passed the course.

(Excerpted and adapted from "Japanese Students Have Surpassed U.S. Teachers," by Bart Ziegler, Associated Press, Greenwich (CT) *Times*, April 23, 1990.)

Arthur N. Curtiss

My dad worked on the railroad. I could see electrification coming, so in college I majored in heavy traction and such. I moved to radio because there were no openings in heavy traction at Westinghouse.

I worked on early radios, musical instruments and front to rear communications on railroad trains. I became supervisor of a small engineering group in 1932. Then I transferred to an Indianapolis operation as Manager in 1938, back to Camden in 1945, transferred to Los Angeles in 1950 as General Manager (First Works Manager) of RCA operations, and finally to Princeton to Davis Sarnoff Research Labs as Vice President, Administration in Research and Engineering in 1961 before retiring in 1971. I stayed with Westinghouse/RCA for my career.

The lows in my workday life were during the depression, worrying about layoffs. I was always given special assignments, mostly as a trouble shooter in engineering.

Once I became a supervisor, I had considerable leeway to get the job done and worked as a team player. There was always competition in the system. I had to overcome the engineering "aura" when I was transferred to the RCA Research Center because of the "esteem" of the Ph.D. scientists. I always made sure communication links were good and was not afraid of getting my hands dirty with the group. Ethically, the biggest problem in a large organization is honesty. As a supervisor, one has to be above reproach.

One of my accomplishments was salvaging a radar program for the Navy in 1950-55 and participating in the Atlas program in 1955-60. My worst failure was being unable to go back to school (if laid off) to become a medical doctor. I was never aware of any discrimination for age, sex or education. The company (bosses) always encouraged my efforts in the company and for my many community activities.

Kenneth R. Jackson

I was fortunate to begin my career in July, 1941, at the Naval Ordinance Laboratory, in Washington, DC. There I associated with some of the best technical brains in the U.S. My boss was Dr. William Brown and his boss was Dr. John Bardeen, who later was awarded two Nobel Prizes in physics—one for his contribution to the development of the transistor.

We did research in the field of magnetics as it applied to degaussing of ships and the development of torpedo exploder mechanisms. I worked two and a half years as a civilian, and two years as an ensign for the same civilian boss.

In January, 1946, I was the twelfth person hired in the new missile department at North American Aviation (now Rockwell International). The department grew to encompass eighty-five thousand employees, but I resigned in five years when it had only five thousand employees. I felt that it was too large.

I was the Autopilot Project Engineer in charge of developing autopilots for North American's first and only two missiles. I was co-inventor of two hydraulic servo valves that were successfully flown on the missiles.

In 1951, I joined J.B. Rea Company as Chief Engineer. We had about twenty engineers and did consulting work (mostly for the Air Force) on the dynamics of aircraft autopilot systems. Digital computers were just being considered at the time. One of our major studies was to determine if the computers were fast enough (at that time) to operate in an aircraft autopilot system.

Rea would not offer us any stock in his company, so the four top engineers and I formed a competing company—Dynalysis Development Labs., Incorporated. I was President and Chief Engineer and too dumb to realize that a consulting company with five sharp engineers could not make it without a sales personality in top management. We sold out in less than five years.

I became Chief Engineer at Waugh Engineering Company which was making turbine flow sensors. We got some companion electronic products going, and invented and patented a turbine type mass flow meter during the two years I was there.

Next, I went to Packard Bell Computer as Assistant General Manager. There my main task was to write proposals for systems including digital computers. Packard Bell had developed one of the first small digital computers. It sold for forty thousand dollars when all other comparable or larger computers cost over one hundred thousand dollars.

My most successful project was selling NASA (Huntsville, Alabama) a computer system for the quality control department to use in checking out missiles and missile equipment. The system included nine computers which accepted digitized information from the test instruments. The tenth computer organized the data and printed it out. NASA patented my idea for an optical sensor system that could tell them how much a fifty foot or so tank or missile was sagging in between supports.

When Packard Bell sold out in 1963, I went to Control Data Corporation (CDC), where I spent a year mostly writing proposals. Then CDC moved the department back to Minneapolis.

In 1964, I joined Leach Corporation as Manager of Engineering of the Relay Division—about nine hundred employees with ten engineers—where I remained until retirement on December 31, 1983.

In 1964 about eighty percent of all electromechanical relays used by the Air Force and by commercial aircraft builders were Leach Relays. I immediately began organizing an electronics group to develop companion electronic products such as time delay relays, and, later, solid state relays and power controllers.

By the time I retired, Leach had two divisions, namely: Relay Division and Control Products (CP) Division. I was Vice President of Engineering and was responsible for both division engineering departments. (I was also acting head of the CP Division engineering department—about fifty employees.) About fifty percent of Leach's sales is now electronic products.

At Leach I worked with the engineers on new products for the CP Division. (I received a patent on an ac power controller.) Mostly I wrote proposals and represented Leach twice a year at the SAE (Society of Automotive Engineers) Power Controller Meetings. I was the SAE Committee Chairman for two years before retiring.

In 1984, my wife and I started K. R. Jackson, Inc., and I consulted at Leach and Royce Electronics until the business petered out. We disbanded the corporation in December of 1990.

At Royce I worked on an electronic ballast for fluorescent lighting. It did not get into production and I am afraid that the company is now defunct.

Max W. Kuypers

I joined IBM in their startup effort for a process control computer system. I spent seven years with them as a project engineer on several "first time" applications of a computer for the control of an industrial process. The project—which changed my professional objectives and provided me with the stimulation and challenges for the next twenty odd years—was adapting a 630 computer to a relay supervisory control system for automatic production of a pumped well oil field in West Texas.

I found that while I could work at a desk either doing circuit or system design or managing a project, I was most happy on my own out in the field. I enjoyed investigating potential applications for remote computer control which, in the 60s, was still in its infancy.

From 1969 until 1988, I worked for nine different firms always involved in the application of supervisory control systems. My position was generally in the design of a system to meet the unique requirements of the specific application. In general, the applications involved oil/gas field production and treating, and the pipelines. My travels took me from the tip of South America to Angola, Libya, the Middle East, former Eastern Block countries, the North Sea, and China, explaining how computer control would enable operating a hot oil pipeline more efficiently.

The project, which stands out in my mind and gave me a sense of accomplishment, was the design, fabrication, installation and checkout of a system which automated the oil production on an unattended platform in the Strait of Magellan off the tip of Chile. The State Oil Company of Chile was losing production because bad weather forced the evacuation of the platform and shutdown of production almost sixty percent of the time. Several large firms had been approached for bids to provide a radio remote control system for the platform. All the firms declined to bid including my employer who also wanted to no-bid. I convinced them to let me go to Chile to investigate the application.

The end result of my trip was a successful bid and my assignment as manager of a project which also involved two sub-contractors for the radio and pneumatic safety shutdown equipment. The installation was carried out in three phases due to bad weather. Once the installation was finished, the acceptance tests were carried out without a hitch.

Using a redundant UHF radio, the client was able to control all functions of multi-well production, separation and pumping to shore with the platform unattended. Included were all the safety shutdown facilities including vibration detection for damage due to waves, ship collision or loss of radio communication.

Elias Weinberger

I found myself, at the age of forty-four, beginning to worry about retirement, especially with sixteen years of U.S. government employment under my belt. I made some calls to Washington, DC, and found that the Defense Communications Agency (DCA) was looking to hire a new Research and Development Chief for the Planning Division. I sent my application in.

After a very interesting interview, I was told that the position was mine. Even though this was a "supergrade" job, it was still quite a cut in salary from the job in Ohio. Considering the other factors, however, the reduction in salary turned out to be very minor. I spent the next five happy years at DCA.

I then accepted a request from the Navy Department's Naval Electronic Systems Command (NAVELEX) to return to the Navy to take over the design of the communication system for the Trident Submarine. I acted for the Chief of NAVELEX on the staff of Admiral Lyons, who was the project officer for the Trident Submarine. The Trident Submarine communication system is now aboard the submarine in all its glory. It was one of the few triumphs of my career.

Jack Staller

During the 70s, I became an entrepreneur. With the former president of an early printed circuit company, I started a company to develop and produce low cost computer aided design tools for designing printed circuit boards. During the development, we became involved with the Digital Equipment Company (DEC). We licensed the hardware and software technology for a computer controlled multi-station semi-automatic back panel wire wrapping system. This had been developed for DEC's own manufacturing, and several hundred stations were in operation in DEC plants in the U.S. and Canada.

At this time, venture capital was available for any high technology. The equipment was developed commercially and marketed in the U.S. and Europe. A wire wrapping service bureau was established, and eventually spun off to become the largest wire wrap service bureau in the world. I served as President and Chairman of the Board. Unfortunately, the software costs far exceeded our budget costs. We ran into a cash bind and were forced to merge with an associate company of GENRAD.

During the eighties, I formed a consulting firm which has become dedicated to supporting the insurance industry. The firm provided support in the recovery and adjustment of high technology insurance claims. The initial thrust was computers and computer peripheral equipment but it has been expanded to cover a variety of high-technology and some low-technology equipment. We have worked on particle accelerators, CAT (computerized axial tomography) scanners, scanners, numerically controlled machine tools, computer aided design systems and such.

In March, 1988, I sold the consulting business and became a volunteer business counselor with SCORE (Service Corps of Retired Executives). SCORE is operated by the U.S. Small Business Administration.

J. Rennie Whitehead

As a result of the link with RCA Victor, I was invited to join them as Director of Research, to create their first Canadian Research Laboratories, in Montreal. In the ten years from 1955 to 1965, the labs grew steadily to include divisions on wave propagation, semiconductor devices, systems, lasers and space.

We designed the first 400MHz transponder for the Alouette topside sounder satellite. We took over from the government the engineering design, construction and test of Alouette II and the ISIS series of satellites. We had a very friendly relationship with RCA in Princeton, which was headed at the time by Jim Hillier, another Canadian.

We competed freely with RCA for U.S. military contracts although it was against corporate rules. However, it was my policy never to compete unless I knew we could win. Ultimately, the corporation had more sense than to criticize a winner. Indeed, RCA Victor turned out to be an excellent environment in which to achieve innovative results. I was given all the freedom I wanted throughout my ten years there—I ran the labs the way I wanted them with only minimal constraints from the Canadian management and RCA International.

G. B. Welbourne

After ten years with two major engineering constructors (Stone and Webster Engineering of Boston, Massachusetts, and Bechtel Corporation of San Francisco) and registration as a professional electrical engineer by examination in two states, it was my plan to start my own company. Twenty years of being my "own boss" was rewarding to my ego when things went well but very lonely without a "home office" to call when I was in trouble.

After a few years of just being able to pay the help, the financial rewards improved greatly. We found that there was a need for a small company that could provide electrical engineering, design and complete installation. Many of our clients have been with us for over twenty years.

The most disturbing problem I faced as retirement age approached (and passed in my case) was to find someone with an engineering degree willing to endure the trials and dirty hands of practical "hands on" field work. One very capable and motivated engineer we employed was persuaded by his wife to go back to his prior design office work. This was because he had to occasionally drive a service truck to his residence and sometimes came home with soiled clothes.

Joe Zauchner

In 1956, I got a job with IBM in Poughkeepsie, New York. This was the best thing that ever happened to me. I started as an electronics technician. After one year, I was promoted to lab specialist; in 1964, to professional engineer; and in 1967, to staff engineer. All the time I worked for IBM, I was very involved with measurements and measurement techniques. I also worked on logic and memory circuit design, but measurements became my specialty.

At IBM it was possible, though not always easy, to work up to professional status without a college degree. I took many of their evening courses. Performance at work, however, was the critical parameter. I had the good fortune to have good managers who were not afraid to give me meaningful assignments.

Not all managers were willing to do this and so I found myself at odds with some of them. Fortunately, IBM policy afforded good protection against incompetent managers and my stay with them never lasted too long. They were probably just as glad to get rid of me as I was of them. Once established, I worked mostly with Ph.D.'s whom I found easy to work with. The time I worked at IBM was easily the best part of my working years.

Warren L. Braun

Past experience paid real dividends and, to my surprise, resulted in some very interesting tasks. One example, the development of a new subscriber located automatic telephone connector. This was used to implement a nationwide talk show, "Night Call."

This program had been based in Detroit, but was failing miserably due to the poor telephone quality of the caller circuits. I developed an apparatus to make this program possible and was awarded U.S. Patent #355190 on the device/system; however, during the eighteen months of system development, the program funding ceased, and the project was put on the shelf. We affectionately called the device, the "GAD" machine, since it did so many things at the same time.

One Sunday, I was relaxing at the Homestead Resort when I received a call from Nelson Price, director of the project. He said, "Warren, you know Dr. King has been assassinated and the inner cities are burning in Washington, Baltimore, and elsewhere. We need to set up the 'GAD' machine in New York City to feed a network of stations across the country to try to cool off the ghettos."

Twelve hours later, I arrived at the hastily assembled production facility in Harlem, New York. I was greeted by a congregation of top flight black leaders from across the country. Thirty-six hours later, after the initial panic, the "GAD" machine was working and a new concept was born.

What was unique about this concept was the ability to have the host in one location, a guest quite literally anywhere else in the world, both of whom could speak simultaneously to the callers, i.e., a truly worldwide conference call program. One of the last programs aired with the host in New York, the guest in Moscow, and callers from across the U.S.

Time called it the "cool hotline" with the magazine cover featuring the host, Dell Shields. *Newsweek* also carried a full page feature on the same program. Sometime later, I received an award for the development of the system.

By 1972, the cable television service arm of the firm had grown to twenty-two employees. I established it as a separate corporation and, with the help of my wife and daughter, named it ComSonics. I kept the consulting firm under my name. My wife was always a strong steady force in the business.

The first two years of the firm were tenuous at best. I took no salary from the company and worked ever harder to establish its place in the market. I also knew I couldn't pay much for the talent I needed and saw the need to develop a participatory atmosphere.

I had been following the writings of Dr. Louis Kelso on "two factor" economic theory, which in 1974 was enacted into law by the very wise guidance of Senator Russell Long, then Chairman of the Senate Finance Committee. Upon enactment of the ESOP (Employee Stock Ownership Plan) legislation, ComSonics became one of the first ESOP companies. Today ComSonics stands as a one hundred percent employee owned firm.

There were some interesting moments in the development of the firm, particularly in 1975 when we were retained to study the future of the electronics industry over the next twenty-five years. The finished study was remarkable in that its findings remain reasonably accurate even today. Since we used a number of consultants, a single copy of the report, including the reference materials, weighed forty-two pounds. Fortunately, the findings and summaries were contained in a document of only one hundred and thirty-six pages. For those reading this, however, the results of the study remain confidential and cannot be disclosed. But perhaps someday it will be available for perusal in a historical sense.

When I left the firm I placed the remainder of my stock in trust for the employees through the ESOP. The employees have brought to the firm a wide variety of skills to bear in some really tough problem solving scenarios, and I am very proud of their results. For instance, they have developed unique proprietary leakage detection apparatus and an automatic level measurement apparatus for the cable industry. Both systems are widely used in the cable TV field today. I look on the firm's one hundred and seventy-three employee owners as real partners in an ongoing trek into the future of communications

Stephen H. Frishauf

In 1941, I was hired by the (then independent) Electric AutoLite Company. Upon reporting for work, the grizzled chief engineer gave me an introduction I will never forget: he said he expected me to make mistakes and requested that I not to make such serious ones that the plant would burn down. He emphasized that if I were to make the same mistake twice, I would be automatically fired. Not a bad philosophy!

Shortly after Pearl Harbor, the plant became engaged in defense work and I was let go since I was not a U.S. citizen. I was turned down when I tried to volunteer for the Armed Forces.

Believing that I would find work in the New York area, with its large civilian work base, I came to New York City. When I reported to the New York Unemployment Office, I was asked just one question: "What is a catenary?"

An employer had specified that question and had provided the correct answer to the interviewer—requesting that the first person who could give the answer be referred. Having had an interest in eletrical traction I, of course, knew the answer. I was immediately hired by an engineering firm to do design work for the catenary system of the Pennsylvania Railroad.

After World War II, I somehow slopped into the patent field and became a patent attorney. This has to be one of the most fascinating professions because one always works at the edge of the new and, often, unexpected—being exposed to the latest technological developments. This old experimenter is forever curious—but always tests first for live wires and grounds!

Howard O. Lorenzen

After graduation from high school, I worked for the Iowa State Highway Commission surveying for the paved roads that lifted Iowa out of the mud. During a visit to Iowa State University in Ames, Iowa, for an Amateur Radio Convention, I observed the University's nice laboratory set up and nice engineering facilities so I decided to apply. I took electrical engineering courses since they did not have courses in radio engineering at that time.

After graduation in 1935, I went to work for Colonial Radio in Buffalo, New York. It was during the depression and I was the first graduate from the electrical engineering curriculum to land a job. I was a radio receiver design engineer. When Zenith Radio in Chicago advertised in the *IRE* for engineers, I took a job in their laboratory as a design engineer for radio receivers.

The next year, a friend from Buffalo wrote me that the Naval Research Laboratory in Washington, DC, had openings for experienced radio design engineers. I applied for a job there and was accepted. I started designing UHF receivers for them and worked on some of the first radar receivers. During the war, I got involved in designing countermeasures for the various German electronic controlled devices.

When the war was over, I organized the people in my group to form the Electronic Warfare Branch. We were very active in developing countermeasures during the war in Vietnam and Korea. The success of our devices provided us with excellent support from the Navy operating forces. Eventually, I was able to raise the Branch to Division status. The Electronic Warfare Division had the best fiscal support of any Division at the Laboratory. We were bringing in about fifty percent of the Laboratory's fiscal support.

Winthrop Leeds

My boss, John B. MacNeill, suggested that I couldn't do better than to attend his alma mater, the Massachusetts Institute of Technology for graduate study.

In 1961, I was given the responsibility of managing the entire power circuit breaker engineering department. However, I found that handling personnel problems, union disputes, and budgetary limitations was not my "cup of tea." I was happier two years later when I was put in charge of a new section—New Product Development. This made it possible for me to supervise the group of engineers that produced the record-breaking 500 kw SF_6 circuit breakers, which I consider my highest achievement. Throughout my career until his retirement, I considered J. B. MacNeill to have been my mentor.

The climax of my career was the application of the gas SF_6 (sulfur hexafluoride) to high voltage switchgear, including the first 500 kw circuit breakers put into service in the United States. It all started with our attempts to answer customers' urgings to find a non-flammable substitute for oil in the design of high power circuit breakers.

Mr. H. J. Lingal, our then Power Circuit Breaker Engineering Manager, suggested that our long range development group systematically test various gases of known high dielectric strength that might possibly do the job. Two of my engineers, Dr. T.E. Browne, Jr., and Albert Strom, carried out such a series of arc rupturing tests.

When they got to SF_6, they were astonished to find that it had remarkable arc interrupting ability. It was first tried out with low-power devices, such as load-break disconnecting switches for high voltage, and circuit breakers for large capacitor bank switching. Success with this equipment led to higher power circuit breaker applications. Plain break interrupting ability was dramatically improved by using a piston to pump the SF_6 gas through arcs drawn at separating contacts. Eventually, the highest power breakers were successfully developed using a compressed gas system that would release a very strong blast of gas for effective high power arc extinction.

Naturally, this development took quite a number of years, studying the effect of arcing on the gas, the application of filters and dryers, selecting suitable insulating materials and so on. I followed this work closely and spent considerable time with our sales people and customers in educational efforts to acquaint them with the advantages of this unfamiliar medium, SF_6. Bob Lawrence reported to me several years later that in a meeting considering my nomination for the Lamme Medal, the chairman, Mr. A.C. Monteith, referred to me as "Mr. SF_6"!

As time goes on, the advantages of using SF_6 in place of oil have become so apparent and used so widely that it has become an industry standard. However, it appears that the pioneering efforts of Westinghouse in introducing the use of SF_6 in arc interrupting devices have been largely forgotten. Nevertheless, the fundamental Lingal, Browne, and Strom patent #2,757,261 issued on 7-31-56 with a filing date of 7-19-51 gives a complete story of the earliest experiments switching arcs in an atmosphere of SF_6.

Yardley Beers

Suddenly I, who had never hired an employee, made out a budget request, or set up a teaching schedule, was made the acting chairman of the department. A few weeks later, an old victorian residence, Butler Hall, which had served the department as its principal office building and research laboratory, was gutted by fire and had to be evacuated. That started the most critical year of my life.

Not only did I have to learn the ordinary duties of department chairman, but I had to supervise the placement of department members in temporary quarters. Once during a hurricane, the roof blew off of one of these buildings.

For four years, I presided over the department. After the fire, the instrument shop was left in shambles and both instrument makers promptly resigned.

One of my first actions was to hire an elderly mechanic, George Timmerman. It was fortunate that I was too naive to realize that others had discriminated against him because of age. He did both jobs, restored order to the shop, and met all requests for building apparatuses.

During the four years I was acting chair, there was not a single resignation of the professorial staff, and, except for possibly a day or two right after the fire, we held all our classes. Some of the research projects were delayed because of the lack of suitable space, but ultimately the space was obtained and projects went forward. I look back on my work as acting chair with complete satisfaction: every task I performed was absolutely necessary, and my work was appreciated by my colleagues.

In 1961, I resigned to take a position with the National Bureau of Standards in Boulder, Colorado. Once I heard Norman Ramsey cite a number of examples of people who had served as acting heads of organizations. They had said that they did not want the head position on a permanent basis, but yet all resigned when someone else was appointed. He can add my name to this list. Event though, I still say that I did not want the chair on a temporary or permanent basis. Yet the prospect of further work at NYU became very uninteresting.

I believed that if I had to go through the trouble of moving, I should move at least a thousand miles. Thus, my family and I moved to Boulder, Colorado.

Marvin Udevitz

Shortly after my arrival in Cheyenne (my original home town), I accepted a position as a Project Engineer with Land-Air Incorporated. I soon lost interest in that job. Luckily, this same company had another division, located at the White Sands/Holloman Missile Range in New Mexico, where they were engaged in design, installation, maintenance and operation of both ground telemetry and missile tracking systems. Thus in the fall of 1954, I took over as Lead Engineer for the MIRAN ground tracking system.

MIRAN, which stood for Missile Ranging and Navigation, was one of the two major ground tracking systems on the range at that time. It had been developed by Oklahoma A&M and delivered to the range in a supposedly "operational" state. But operational in those days didn't mean what it does today. The system was built around a 2-GEDA computer. This had a zillion racks and ten zillion vacuum tubes, mechanical resolvers, and all the other devices of the day which either didn't work or required constant attention and calibration.

And what did all this equipment do? It solved a simple set of simultaneous equations with three variables from a possible six radar range sources. The computer ran at the fabulous rate (in those days) of 10 khertz. One could do more with a $69.95 hand-held programmable calculator today...but no one could have been more proud of our achievements considering the state of the art.

We somehow kept all those vacuum tubes and mechanical devices operating while we tracked Aerobees, Radio Plane drones, Sidewinders, Matadors and anything else which showed up on range. I believe all the people involved in those days would readily admit that we were really groping around in trying to get things done. It seemed that no matter what we wanted to do, neither the resources nor the techniques existed. We had to invent everything. In truth, however, the early days in the missile business, whether one was involved on the ground side or the airborne side, were the most fun. Later on, when things got more disciplined and a lot of emphasis was put on quality performance, we had a lot more successful missions, but it was a lot less fun.

The other major event in my Martin Marietta history involved the Viking program in 1976. Although I was not assigned to the Viking Program until relatively late, it was in time to be designated as the Systems Project Engineer for the final test phases at Denver and the launch operations at Cape Canaveral. For me, it was a real double header. I had been part of the design team for the Titan/Centaur which was the launch vehicle. Now I found myself associated with payload, itself.

Following the successful launch of both the Viking spacecraft, I was then assigned to the Mission Operations team at the Jet Propulsion Laboratories (JPL) in Pasadena. My specific job was leader of the Lander System Analysis Team (LSAT). My team actually was quite small but it was composed of super experts on every subsystem on the lander. As the world knows, both landers achieved successful soft landings, each one of which was a small miracle in itself. Being at JPL and witnessing the landings in real time—eyeballing the first pictures from Mars on the video displays in our LSAT operations area—

was perhaps the biggest thrill I have ever had.

I found out early that when you accept (and desire) responsibility, and prove your ability to work with a minimum of supervision, that your superiors are quite willing to put more and more on your plate. This is especially true when they get you at a bargain price. Because I didn't have a degree, I always had to prove my knowledge and ability by demonstration. I also was always lagging on the pay and promotion rewards.

This didn't ever matter as much to me as the opportunity to learn new things and enjoy most of the research and development projects. I also found that I got along well with my fellow engineers and technicians whether I was leading or following.

Samuel Sensiper

With almost a free hand, we designed and had built a wide range of equipment including one of the early calibrated signal generators. Of course, we were interacting with other groups at Sperry as well as visiting the MIT Radiation Laboratory, and reading the reports from other laboratories.

The above activities were during the period of about 1943 to early 1946 and were, on review, one of the most productive periods of my career. Minimal supervision led eventually to the admonition of my boss's boss that I had spent an enormous amount of money (for those days) on developing and building a large supply of test equipment items. But the comment, "Thank God you did because the tube people and systems people wouldn't have had any test equipment otherwise," indicated approval for my having the judgement and daring to proceed. (Or perhaps my supervisors knew all along what my group and I were doing and were allowing youthful exuberance and excitement to be creative and productive.)

I opted for Hughes to work under Van Atta in the antenna department in the radar laboratory, starting in August, 1951. After a short time, I was put in charge of the research section of the antenna department. Aside from performing as the section head, I was able to do some of my own research on slots on circular cylinders. This led to a talk and a paper.

I had begun to realize that I had been delinquent in publishing the results of my work and that for many reasons I had better overcome this fault. Much of my wartime work had been written in company reports which found their way into various repositories for use by other people and groups. But the possible further recognition that might have come from open publication when permitted would have been useful. The fault was mine and I intended to overcome it. Hughes strongly encouraged publication back then.

Although working for Van Atta was an interesting experience, I often found that the people I was responsible for were working under his direction rather than mine. In view of his experience and reputation, I didn't directly mind this except that on more than one occasion I seemed to be the last one to know. I felt this was a bad management procedure and I have since learned that it is not uncommon! Later in my career, I believe I learned how to cope with this situation, but then I was not so mature.

In 1970, the electronics and aerospace industries suffered considerable retraction which continued for several years. TRW was among those considerably affected. The last several months there I was engaged in reducing the employment in the Antenna Laboratory since the available financial support was evaporating. I had never had to do this in any of my previous positions. It was somewhat painful and exhausting. However, I knew this came with the territory of being a manager. Finally it was my turn. I was told that my services as Antenna Laboratory manager were no longer required. I left TRW in the late summer of 1970. I had never been fired from a job before. I decided to determine if I could survive as a consultant or consulting engineer.

After a few phone calls and a few months, I had acquired a client. In the years from 1970 until September 1989—with the exception of one and a half years—I have been self-employed. Several of my clients were companies where I had been employed in previous years. Indeed, I have been employed as a consulting engineer by every company I had ever worked for, on the west coast at least.

After I became a consultant in 1970, it seemed wise to have a more recognizable standing so I acquired a professional engineer's license in California in 1977. The professional license has never been particularly useful in my consulting activities, but I still believe it could be and will perhaps become a recognizable distinction for all individuals who call themselves engineers.

Francis J. Heyden

Around the end of my high school, I decided that I did not want to go off as a "sparks" on an ocean liner, but instead as a Jesuit priest. Learning Latin, Greek, history, chemistry, physics and philosophy, I had some time to work with radios.

A Jesuit superior asked if I would delay my doctoral studies and go to the Jesuit college in Manila of the Philippines to teach college physics. My physics minor qualified me. In August, 1931, I arrived in Manila and took over a class of forty students within two days. The teacher was leaving on the same ship on which I came. Teaching was a new kind of life and I learned more about physics just working with the laboratory experiments.

After five months, another Jesuit superior in the Philippines sent for me, "We did not bring you here to teach physics. We want you to be chief astronomer of the Manila Observatory."

Manila Observatory had started in 1865 when a Jesuit in the college made a forecast of a typhoon. The Spanish government began financing the little observatory which also started astronomy and observations of time for ships in the harbor. A time-ball was dropped every day at noon on the roof. Seismology was also introduced along with magnetism. It was a complete unit for public service in weather: seismology, time and magnetism. Jesuit priests had full charge. In 1898, the U.S. took over the Philippines and decided to keep the weather bureau as it was under the Jesuits.

So I became "Chief Astronomer" within ten days after the appointment by the Jesuit superior. I needed those ten days badly. Because while teaching college physics by day, I worked nights with my predecessor, Charles Deppermann, S.J., learning how to observe time stars, rate the master clocks and transmit the evening time signals by wire to the U.S. Naval radio station at Los Beanos and Cavite, to the Bureau of Posts and to the Philippine Railroad.

An earthquake shock came about every two weeks. It was strong enough to shake the pendulum clocks off their regular error. One clock, a "synchronome" had two pendular in synchronism. The master was in a constant temperature room in a 6 millimeter vacuum. Every thirty seconds it received an impulse from the "slave" pendulum, which moved the clockwork. The impulse released an arm with a tiny jewel that fell against a little wheel on the "master." The position of the wheel timed the return, pulse, to the "slave."

With the "slave" set to run slow, the wheel on the master determined whether or not the "slave" received an impulse from a spring that dropped into position at the end of its swing. The spring shortened the amplitude and sent the pendulum back on its return. This synchronism kept the two pendular in perfect synchronism. But an earthquake could throw out the synchronism enough to drop the jewel in the master so that the wheel would meet it head on! At least once the jewel was broken.

As soon as the tiniest tremor came, even from my bed at night, I would be running for the clock vault to stop the impulses from the "slave." I soon learned how to get the pendular back in synchronism with my finger on the "slave" or with the air valve on the "master." When the tremor had slowed down the amplitude of the pendulum, a jet of air from the valve would give it a push. This saved the trouble of opening the vacuum case to push the pendulum. The former director told me about this just before he left.

The observatory got its radio station. We had an RCA one kilowatt transmitter and an unusual receiver which a telegraph operator turned "on" and "off." I fixed everything. "Signal tracing" was new to me but I kept every piece running.

I was sent later to Washington to work with the U.S. Weather Bureau and work with Georgetown College Observatory. From 1945 to 1946, I drew the daily weather map for the United States and then compared it with the routine one done by the staff. I followed isobars strictly while the staff followed continuity by moving the frontal lines forward. We did a map every three hours.

At the observatory, I worked a bit with the telescopes and waited for the liberation of the Philippines. It came in 1945 but the news brought Charles Deppermann S.J. to Georgetown. He was wanted for his knowledge of weather for the final attack on Japan. Everything was destroyed at Manila Observatory. In the last days, the buildings were burned and their surviving walls were forts for an artillery duel between the U.S. and Japan.

A letter from New York told me I was the Director of the Observatory and to look for some new Jesuits. The former director, Miguel Selga, S.J. had resigned. Charles Deppermann, S.J. wanted to go back into astronomy. He was finishing some papers he had lost while a prisoner at Los Beanos. Bernard Doucette S.J. wanted to do seismology. William Repetti, S.J. resigned. Leo Welch, S.J. resigned. He had a degree in meteorology from MIT. I was lost for a while. Some said to forget the Manila

Observatory. Finally I agreed to turn weather forecasting over to the Philippine Government and keep astronomy, seismology and ionosphere work for the Jesuits.

I began writing war damage claims. For the library, meteorological, seismic, astronomy, magnetism and even the lost buildings. I received enough money to get started.

All this went on, while I stayed at Georgetown University. I started a graduate department of astronomy which turned out some ninety graduates, in some twenty-six years.

There was also the radio station, WGTB. It began with a small six watt transmitter that fed standing waves over the lighting circuits. I got to know the heat tunnels very well as I coupled all of the power transformers into the six watter.

Besides the six watter, I undertook three outside broadcasts. First was a broadcast of "The Mass for Shut-Ins" from the chapel. Second was the "Blue and Gray Show" of variety entertainment by college students every Saturday, that lasted for five years until I gave up. Students were fine but professors who wanted to help were impossible. Third came from the Georgetown University Forum.

When I left Georgetown in 1971 to return to the rebuilt observatory in Manila, Georgetown closed the observatory and gradually stopped all broadcasts including the FM station. I have heard that the six watter is still going. During the FM years, the six watter carried "musak" for study music.

I still have some "Blue and Gray" show tapes and some special broadcasts I did for NBC on "Breakthroughs in Science." These are like the candle stubs that the hunchback of Notre Dame saved as treasures.

George H. Barnes

At any given time, I could tell you what my specialty was, but it has been a moving target. From electromagnetic theory, I moved into the circuit design of special-purpose analog computers. After five years of that, I was hired by Burroughs to join the Circuit Techniques Section, whose charter was to invent design methods for digital circuit designers. When I joined, we were working to replace vacuum tubes with magnetics, for digital applications. I was a co-author of *Digital Applications of Magnetic Devices*, edited by my boss, Al Meyerhoff, and published after magnetics had been mostly obsoleted by transistors. Eventually the magnetics went away.

Starting in 1966, I spent six years on the Illiac IV project. In retrospect, the Illiac IV was only a partial success. It ran at great (for then) raw speed—Harv Lomax' program ran about 60 Mips—but presented a rather gruesome interface to the programmer. I'm a co-author of the paper describing its architecture. The rest are mostly from the University of Illinois.

After Unisys paid me to retire (1981), I went back to electromagnetics, becoming an electromagnetic compatibility (EMC) expert for ORI, Inc. (formerly Operations Research, Inc.).

A common thread in all of this is an ability to solve problems with a little bit of applied mathematics. There have been times when you could say applied mathematics was my specialty. My last job at RMS Technologies is an example. I had to debug the algorithms being used to track radar targets, and then debug the conversion of algorithms into the program. There were glitches at both levels.

At least ninety percent of everything I have worked on was fun. But I have to say the speech-processing work was a notch more enjoyable than anything else. It used things that came out of many different places in my past. Things I didn't know were related until I found myself using them together on this one project—acoustics courses from college, z-transforms from the analog-computer days, and so on. The speech processing also had the gratifying property that heuristic, home-made algorithms often produced a better reproduction of the speech than mathematically respectable methods did.

From '79 to '81, I was the principal investigator of a proposed parallel processor to serve as the engine in NASA Ames' Numerical Aerodynamic Simulator (NAS). I believe we described a machine which like, Illiac IV, was very programmable. Unfortunately not even a small-scale model was built.

New technologies were never a threat. I was often a member of the team that was introducing them. We were going to replace vacuum tubes with magnetics. We helped replace vacuum tubes with transistors.

I've always been good at teaching myself whatever I had to know. I've been lucky to work with knowledgeable people who are willing to teach, not formally usually, but always helpful.

I have not enjoyed some of the changes that were imposed for business reasons. In the mid-60s, Burroughs closed its research center in Paoli, which was in the same building where my job was. Later, the technical library in the same facility was closed. For a while, we kept most of the books in a book room, but even that eventually went. Good engineering is easier when necessary references are on the premises.

Aubrey G. Caplan

The job I took was with the local power company, Diquesne Light Company, Pittsburgh, Pennsylvania. I was hired and trained to be a power salesman in the commercial division. My job was to promote the use of electricity by designing fluorescent lighting, air conditioning, electric cooking for restaurants, miscellaneous electrical uses, and larger service entrances when required by the increased usage. During my time at the power company, I received my professional engineering license.

On this job I learned utility rates, electrical distribution, wiring design, advanced lighting techniques, restaurant layout, air conditioning sizing, trouble shooting, diplomacy, and salesmanship. I designed new lighting for everything from a chicken coop for a farmer who wanted time clocks to fool the chickens, to a bordello owner who didn't want to use very much light. In my job and travels, I met numerous architects, builders, property managers, and supermarket owners; these contacts would prove valuable to me in the years to come.

In 1955, a small shopping center was going into my territory. The plans were so bad that I sat down and redesigned the entire layout in order to get logical bids. After construction got under way, the owner called me into his office. He said, "I understand you redesigned my whole job. Without your help, I would have been in big trouble. What do I owe you?"

I replied that I did this work as part of my job at the power company. Also, I did it for my benefit as well as his and that he owed me nothing. His reply was for me to pickup drawings for six more jobs and this time to send him a bill for time spent on his work after five p.m. He did not ask for a price.

This was the start of my career as an electrical consultant. My patron recommended me to other builders and to architects. From 1955 to 1959, I had two part-time employees working every night and weekends out of my basement. When my income for part-time work exceeded my daytime salary, my accountant said I had to make up my mind whether to quit the power cmpany and go full time on my own, or to turn work away, as I could not do both. In 1959, I said good-bye to the power company and went full time as a Consulting Electrical Engineer.

In the past thirty years, my office has designed over four thousand jobs. Robert, my son, joined the firm. Together we have designed just about every kind of commercial and industrial electrical installation. We designed a summer home in Greece, a food radiation facility in Hawaii, a plutonium laboratory in Japan, a convent-school in the hills of Puerto Rico, and an electric locomotive.

My office has received numerous lighting and wiring design awards. I have had two articles published in *E.C. & M. Magazine* with my picture on its cover, as being typical of the "bread and butter" engineers, who design the everyday routine jobs. I am still actively engaged in the business of electrical design with no intention of an immediate retirement. I have seen my contemporaries "go to pot" when they quit working.

Louisa S. Cook

Seniors (only four in electrical engineering) were interviewed during our last semester at the University of Arizona. I came to work as a junior engineer for the Salt River Project (SRP), 1947-55, in Phoenix, Arizona. After learning the size and layout of the power system, I was responsible for calculating system impedance and preparing data for analog power system studies. Since I enjoyed learning in advance how to recognize accuracy of information, I had always been a welcome member of college lab teams, and continued to fill data acquisition positions after gainful employment in the power field. Electronics instruction at Arizona was not available until veterans of World War II returned and demanded it in 1947.

Upon completion of a few digital computer courses after 1960, I resumed employment with SRP to prepare system planning studies data. I never was permitted to run the studies (on out of town computers). About a year later when given the choice of signing an overextended expense account for my immediate superior or resigning, I immediately resigned. I had never been outside our office or away from my desk!!

I was satisfied with the work I was doing so I never attempted to move into management. This was probably just as well; because the chief engineer during my later years with SRP responded in 1977 to a request for a reference letter as follows: "I am not the best person to contact relative to working wives as I put my foot down at my own home. I don't see how a woman can carry on in the engineering profession and raise a family, too, and do justice to both." By this time I had raised three children—a son who is an electronics engineer, a daughter who is a commercial airline pilot, and a daughter who is an animal scientist working on her pilot's license.

Then in my fifties, just as our youngest entered college, I was hired by Bechtel Power Corporation as a field engineer at the Palo Verde Nuclear Power Plant. (Much to the surprise of that ex-SRP chief engineer when he saw me there one day.) I retired right after I turned sixty since that met Bechtel's retirement requirements. It was a good thing I did, too, because the next year my husband, George, had a bout with cancer from which he completely recovered.

I can't imagine a much finer sequence of events. Our children seem to be following a similar sequence: enjoy career, slow to marry, continue essential work and live happily.

W. Jack Cunningham

My graduate work was almost completed so I was looking for a job for September, 1946. A Harvard graduate student, with an undergraduate electrical engineering degree from Yale University, suggested I explore a faculty position at Yale. I did so and, just before the fall term opened, I agreed to go to Yale as a lecturer in electrical engineering at an annual salary of four thousand dollars.

With the great influx of students and faculty following the war, housing in New Haven was hard to find. Al Conrad, who hired me, found for us the second floor of an old row house. A family of a full professor of physics (back from the Navy) occupied the floor above, and a law student's family on the floor below. Our kitchen had a flimsy metal box that required a man delivering a block of ice each day. The oil burner for a common steam heating system shut off all too often because the available fuel oil was of such poor quality it would fail to ignite. We had no auto, but the streetcar system allowed us to get about New Haven.

Electrical engineering at Yale was making the transition from power machinery to electronics. At the same time I was hired, Herb Reich joined the faculty to boost electronics and John Bower came to start work in servomechanisms. There were so many student in 1946-47 that classes began at eight in the morning, and laboratory sessions were being conducted in the evening. An experiment on resonance was done using an alternator as a voltage source, driven by a variable speed dc motor, both of about ten horsepower capacity. Electronic circuits for experiments were constructed on aluminum baking pans in lieu of hard-to-get metal chassis.

The first few years I was at Yale, I did occasional work for industrial organizations, and from time to time considered shifting out of academia. It was not long, however, before I found myself buying my own cap and gown to take part in the annual graduation ceremonies. I realized that I had made a commitment to the academic life. I remained in the engineering department at Yale for forty-two years, and retired in 1988.

During that time I rotated in and out of most of the positions that can be filled by a faculty member. Many changes took place in the administrative arrangements for Yale engineering. Some worked well; some worked less well. The technology continually changed, from tubes to transistors to silicon chips, from slide rules to calculators to mainframes to personal computers. It was an interesting challenge to keep up to date, and to transfer what was going on into courses that could be taught to students. I taught courses in the areas involving electrical engineering, classical physics, and applied mathematics.

Emil C. Evancich

My next job was with Northern Electric Company who manufactured heating pads and electric blankets. I was head of the manufacturing engineering department. At this plant they also made the heater wire and lead wire for electric blankets.

Offered a job by Wirekraft, I told the owner I would only work for him if I was part owner. We agreed on this and I became the chief engineer for all his plants. We were very successful. I developed many products, cost cutting machines and patents. Just about every home in United States had a product I designed, such as a refrigerator heater wire, ceiling heater wire, heaters for pipes. I also took the Indiana test and became a PE (Professional Engineer). In 1959, I was promoted to president of the company. It was sold in 1964 and I left.

I moved to California and took a job as head of computer stock market research with William O'Neil & Company. I made a computer model of the stock market by using statistical techniques to research investment strategy.

I divorced after twenty-four years of marriage. I left on a two year trip traveling around the world. I met Trudy Hooksma and we got married.

I started Wireflex, an electrical wire manufacturing plant in Chicago. Then I purchased property with a building in Bourbon, Indiana. Wireflex became very successful. We went into drawing and stranding wire from copper rod and compound PVC thermoplastic. The company merged with Burcliff Industries and I sold my interest.

I became chief engineer for Wrap On Wire, which manufactured heaters to prevent pipes from freezing. I set up the plant to manufacture wire and become president of Wrap On Wire.

I left to start Wirepower manufacturers rep organization because of age discrimination. We sold machinery and supplies to the wire industry in the mid west. I was 60 years old and wanted a job. I had been very successful in the wire industry, but no one would give me a job.

I was very lucky to come across the wire and cable industry as a young engineer and to make it my life's work.

William A. Edson

My career is checkered. I have worked as an individual at the bench, as a group manager, and as president of a small company. I have worked for large companies (General Electric and Bell), and one small one—Vidar. Also, I have taught at three major universities and one small college. Each of these positions offered it's own special set of gratifications and frustrations. I believe I was a fairly good teacher and a reasonably good group leader; I was not a successful entrepreneur.

I worked at the Bell Labs from August, 1937 to August, 1941. At that time, I was assigned the task of developing a high-gain IF amplifier with a bandwidth of 10 Mhz centered at 60 Mhz, to meet the needs of upcoming X and K band radar systems. The best available tube was the RCA 1852, later designated 6AC7. Its bakelite base and iron envelope were poorly suited to the application, but the all-glass miniatures had not yet been developed. Naturally, this work was highly secret. A vivid memory of that time is sitting silently (and glumly) through an IRE meeting in New York at which an RCA engineer, working on television, pointed out that it was impossible to achieve the values of gain and bandwidth that I had already exceeded.

Seeking greener pastures, Al Ryan, Ed Proctor and I left GE and founded the Electromagnetic Technology Corporation (EMTECH) in Palo Alto, California. Alex Poinatoff was our landlord and Ed Boshell, a prominent Wall Street figure, provided some funds and much organizational know how. I was Director of Research, but things did not go well, and after a year I was promoted to president. We built high power filters, microwave windows, channelizing filters, and microwave gain equalizers to flatten the response of traveling wave tubes. We were never really profitable and went through various reorganizations, ending as a subsidiary of American Electronics Labs (AEL), near Philadelphia, Pennsylvania. (In the spring of 1970, AEL decided to close our shop and move the operation back east.)

I didn't want to move, and found a job at the Viacom division of Vidar Corporation in Mountain View, headed by my friend, Vernon Anderson. There I learned about the T-1 telephone carrier system and tried to develop a cheap system for sending it over a microwave link. Unfortunately, at about the same time, Viacom was acquired by a (larger) Saint Louis corporation that already had such a link. In the face of this reality, Verne, very generously, gave me three months to find another job.

Early 1971 was not a good time to find a job, and I really beat the bushes, without much encouragement. Then in February, Sally and I were invited to dinner at the home of Ed and Elizabeth Proctor, old friends from GE and EMTECH days. Also present were Ray Leadabrand, one of my former students at Stanford, and his wife Millie. Upon hearing of my troubles, Ray offered me a job in the Radio Physics Lab of SRI International, formerly Stanford Research Institute. I am deeply grateful to all of them for this kindness, and remained at SRI for over eighteen years.

At age sixty-five, I was obliged to give up my managerial and supervisory duties, but was allowed to continue as an individual contributor and project leader. I very gradually slowed down and ceased to lead projects. Now I work about twenty hours a week on several problems concerning electromagnetic scattering and antenna near-field distributions. In this way, I keep active and avoid the problems that are encountered by some engineers who retire abruptly.

Chapter six

On the job

Donald Shover

I think the most interesting time of my engineering life was the period of 1943 to 1945 when I was the instrumentation engineer for a group of radio-chemists at Clinton Laboratories in Oak Ridge, Tennessee. I was responsible for providing all of the radiation measurement instrumentation for the group and maintaining and supervising its use.

New radio isotopes were being identified and characterized frequently. I saw the first macroscopic sample of plutonium when our building staff was solemnly marched past a large test tube of a dark purple liquid.

When I finally got dangerous with vacuum tube circuit design, the transistor appeared and had so many advantages that it was indispensable. When I finally got comfortable with transistor circuit models, integrated circuits appeared and design became more a matter of system design than circuit design. Of course, this was all done in the analog world. Now came the design of measurement techniques using sampling and digitizing techniques. I loved it all and had the rare opportunity to play with it and be in on the first applications of computer chips. I think what most impressed me about computer chips was that they were so inexpensive (cost, space, and power wise) that you could afford to waste their capabilities. I asked for and was given permission to go to all of the pertinent company divisions to tell them what these devices could do. Also, I told them how they could be applied to that division's operations.

Most of my career was involved in research and development for the government. This usually involved answering RFPs (Request for Proposals) with formal and rigidly formatted design concepts. In order to have any chance of winning a contract, you had to be unrealistically optimistic regarding the anticipated cost, schedule and the technical chances for success. I always had guilty feelings about a bid that I had worked on, but that's what the government demands and that's what you have to do to get a job. The only mitigating factor is that I was involved in a number of highly successful jobs that were procured through this unrealistic process.

Theodore Schroeder

The interconnection of power systems, already accomplished somewhat in the eastern part of the country, was now being studied more extensively on the west coast, and in the northwest, the midwest, and the south. Many things were learned by those of us, who *knew* all about them, when checking them out on the network analyzer.

One memorable instance occurred when we "tied together" four groups of systems (the south central, the south, the TVA system and some of the northeast). We set scheduled interchanges of power for each of the four main tie lines but to our surprise at least one of the lines would not carry the load per schedule if the other three did. After thinking for a while there was something wrong with the analyzer, we realized that we were trying to violate a version of Ohm's law—the impedances in the particular network set-up had governed the loading of that tie-line not our schedule.

The infamous "Northeast Blackout" of November 19, 1965, did not reach the midwest but it banished some complacency on the part of utility operating groups generally. I recall speculating with Charles Concordia, power system expert of General Electric, at the 1963 summer meeting of the AIEE that we could someday have a major collapse of the interconnected system. The actual event led to the installation of "black start" generation, the addition of under-frequency load shedding relays and the establishment of regional coordination among groups of utilities [IP is in "MAIN" (Midwest Area Interconnected Network)].

In the late '60s, nationwide concern about our polluting the environment became so strong that various state "pollution control boards" were required to adopt more stringent regulations governing water and air pollution. They were soon reporting to the EPA (Environmental Protection Agency), the federal organization set up by Congress.

Numerous proposals came forward from new and also established consulting groups and manufacturers for controlling SO_2 (sulfur dioxide) and NO_2 (nitrogen dioxide) emissions. We carefully reviewed these and found most of them impractical and all of them very costly.

In one case, the company entered a joint venture with the EPA and a chemical company. Under the guidance of the latter, we were to insert a complete sulfuric acid making plant in the stack gas passage of one unit in the plant located in a "major metropolitan area" (having the lowest permissible levels of SO_2 emissions). A special catalyst was used to facilitate the formation of sulfuric acid from the SO_2 in the stack gas.

Each party in this effort shared an eight million dollar expenditure for the equipment and its installation, plus an extensive monitoring program by an expert consulting firm. This facility was unsuccessful because, after considerable operation of the generating unit under normal load levels, the acid produced at the lower loads was quite dilute and, therefore, extremely corrosive. The equipment corroded so extensively and rapidly that the facility was judged impractical.

Charles R. Smith

Early in my career as a development engineer of new products, we had a great deal of freedom in developing the products, and much prestige and confidence in our role within the company and its future. (Some of our most significant products that made us number one in the marketplace and gave us our most substantial profits were developed during this period. Some are still in production today.)

However, this gradually changed starting in the late '60s. The "number crunchers" wielded increasing influence on engineering activities as to what, when, and time (costs) of new product developments. This new approach was, I believe, justifiable for our company's survival and to meet the investment community's demands.

Frank R. Stansel

It has been said that the road to progress is lined with junk heaps. And all too often the stories behind those junk heaps are forgotten. May I present the story about a portion of a junk heap I had a part in making?

In 1927, the first commercial transatlantic telephone circuit was established using a radio circuit operating at 60 kilohertz. The service was so successful that AT&T began to look for ways to increase the service. A frequency assignment of 68 kilohertz was available so plans were begun to add a second low frequency channel.

Our group in the Bell Telephone Laboratories, at the field station at Whippany, New Jersey, had finished designing the first 50 kilowatt broadcast transmitter which was installed at station WLW. We were asked by AT&T to design an improved radio transmitter for this new service. It was proposed to establish a new radio station in Maine with two radio transmitters using this new design. One transmitter would serve the new 68 kilohertz channel, the second would replace the 60 kilohertz transmitter then in service at the RCA station at Rocky Point on Long Island.

Our group constructed a model of the proposed transmitter. The final stage, which could deliver 300 kilowatts at 68 kilohertz, initially consisted of about twenty large water cooled vacuum tubes. Later, a double ended water cooled tube capable of dissipating 100 kilowatts became available. The last stage in the new transmitter was changed to use six of these tubes in push-pull.

Both transmitters were to use a common antenna. This antenna, instead of being the conventional, low frequency type on four hundred foot towers as at Rockey Point, was to be a wave antenna type. This type consisted essentially of an array of long lines. Such antennas had been used successfully for the reception of radio signals but had never been used for transmission.

There was uncertainty about whether larger currents used in a transmitting antenna caused some distortion due to possible nonlinear elements in the ground. To clear up this point, AT&T built a test antenna in Maine just across the Pennobscott River from the University of Maine campus at Orano. The antenna was fifteen miles long and consisted of a single wire mounted on thirty foot telephone poles. To measure distortion created in an element of a single sideband transmitter, such as this one, two frequencies are applied and the third order modulation products (of the type 2A-B) are measured. We needed to energize this antenna simultaneously with two frequencies.

For these tests, we built a queer radio transmitter. We were authorized to operate on any frequency between 50 kilohertz and 75 kilohertz except for certain reserved frequencies, one of which was 68 kilohertz. (Sixty-eight kilohertz had been reserved for the new AT&T station.) Our test transmitter thus contained two channels and delivered to the antenna 12 kilowatts at 67 kilohertz and simultaneously 12 kilowatts at 69 kilohertz. The final stage of each channel was a water cooled vacuum tube. The outputs of these tubes were combined and then, to insure an absolutely clean test signal, a band pass wave filter was installed between the transmitter and the antenna. As this filter had to pass 24 kilowatts of power, I doubt there has ever been another wave filter as large as this one built.

Why did this project never see commercial service and eventually land in the junk heap? Economic and political conditions in Europe at that time were in turmoil. The British Post Office found they could not meet the original proposed service date and so the project had to be postponed a couple of times. Then with the commencement of hostilities of WWII, the project was indefinitely postponed.

After the war, it became quite evident that the low frequency radio band with its limited band width could never meet the need for transatlantic communication which was growing larger each year. So the use of low frequency radio was replaced by a newer technique, the cable and later satellite transmission. However, it wasn't a complete loss. Personally, I learned a great deal on this project as, I am sure, the others in our group did. And some of the ideas explored were used in other applications later.

E. E. Thompson

I joined Arkansas Power and Light Company (AP&L) in 1949. I worked as a distribution engineer in the Central Division in Pine Bluff, Arkansas until I was recalled to active duty in the Air Force in 1951 during the Korean Conflict. I returned to Pine Bluff and AP&L in 1953 and subsequently was promoted to Senior Distribution Engineer in the Central Division. I was transferred to AP&L's General Office Engineering Department in 1958 and held positions of Distribution Planning Engineer and Planning Engineer before being promoted in 1978 to the position of Manager Transmission & Power Supply Planning. I retired from AP&L in 1988. I always worked as a company team member.

I enjoyed engineering planning work from the beginning—not only the planning of an electric system but also the interface with engineering and management employees in other departments and divisions within the company. I also enjoyed the joint planning with system planners for other interconnected power systems, and serving as an expert witness before state and federal agencies and regulatory authorities. The executive management of the company was very supportive of Long Range Planning and System Planning activities.

Walter Schweiss

My career coincided with my training as a project engineer at the Philco Corporation. The responsibilities required leading a group of engineers and technicians to provide prototype radio models for each production year of the Mercury automobile: 1956, 1957, 1958 and 1959. These radios were designed as hybrid models, i.e. applying transistors and vacuum tubes to the circuit design. For economic reasons, only power circuits used transistors. The radio signal circuits used vacuum tubes.

The Philco Corporation formed a Solid State Study Group under the directorship of Mr. William Forster from Harvard University. At the group meetings, assignments were given to the various participants. As a regular member, it was my responsibility to review Dr. William Schockley's Bell Laboratories research notebook on solid state circuit theory. My assignment was under the guidance of Dr. James B. Angell and Mr. F.P. Keiper. Their theory of circuit applications of surface barrier transistors was developed and I was able to fabricate and test these circuits successfully. This resulted in producing the first all transistorized radio for the 1956 Chrysler Imperial automobile.

Philco's release of the all transistor automobile radio for the 1956 Chrysler Imperial (a quantity of one thousand radios) caused quite a stir with General Motors (GM). Prior to Philco's delivery of the radio, it was publicized in a full page announcement in *The New York Times*. There was a rebuttal by the GM company vice president stating that Philco was premature by two years and could not accomplish this goal. Since the Chrysler Corporation had confidence in Philco, the Chrysler president offered to deliver the new Chrysler Imperial automobile to GM including the transistor radio prior to production if GM was willing to pay the full cost of the automobile. As a result, Philco delivered and GM subsequently released their transistor radio for their Cadillac model car the following year.

H. R. Weiss

The highlight of my career at General Electric (GE) was my work for the space program. The group that I worked for built and designed test equipment for their radio guidance system and I had part in that. Several trips to Cape Canaveral and my witness of the first launch of a Venus probe are unforgettable to me. At GE's urging, I published several articles in electronic magazines. GE also showed enormous interest in having their engineers pass the Engineering License test; they gave courses on the subjects to facilitate passing of the exam.

As an engineer, when given an assignment, you are never certain if a solution exists; and if it does, that you are the proper person to come up with that solution; and, even if you are that person, you are never sure that your boss won't change his mind and give the job to someone else in midstream.

I never had the opportunity to go into management, although I believe, that had I stayed with GE in Utica rather than move to Valley Forge (where I was the new kid on the block) it would have happened. I had always hoped to be a manager over an engineering group, knowing my strong points in that field, but, maybe others saw me in a different light.

Price Wickersham

The best library backup. At a place like Midwest Research Institute, customers would come in and expect you to pretty much be an "expert" on a wide range of applications. My engineers and I would do our best to put up a facade of understanding at initial meetings until we could hasten to the Linda Hall Library a few blocks away in Kansas City. That, and perhaps thc John Crerar library in Chicago, are in my opinion, probably the two best technical libraries in the country, supporting clients all over the world. (I say that, having gone to several of the big west and east coast university libraries.) We'd go to Linda Hall and give ourselves a crash course on whatever subject. We could count on Linda Hall having what we needed. When you aren't too smart that makes a real difference.

The course of reactionaries. To me, "reactionary" is a dirty word. I always told my guys they could call me damn near anything they wanted to, but never call me a "reactionary." The engineering profession in its innovative pursuits is confronted so often by reactionaries who will not readily accept new ideas that have a modicum of risk.

Two examples: The lawyer/accountant team was a major adversary. We undertook the development of a minicomputer in 1961 as sort of a bootleg project (before disk memories, cheap RAM/ROM and the PDP/8) and had our chain pulled after three to four months of work.

The reason from the lawyer/accountant adversaries: "IBM isn't doing this!" And then in 1969, we made a fairly detailed proposal for an in-circuit test system development. At that time, there were none on the market, a market that in the seventies grew to several hundred million. No funds were authorized for this new direction, but funds were authorized for more routine extensions of what we'd been doing.

In my projects, almost all had significant unknown segments. Stuff that hadn't been done before (at least by me). So it's been a succession of pre-ball game commitments on time and money, euphoria and fun for the first few innings, some concern in middle innings, and in many cases, sweat and stomach-churning coming into the ninth. Some ball games go into extra innings or are just plain lost, either result with a significant flurry of recriminations.

One example: we had committed to deliver a very complicated $1.5 million system to a big Japanese company. It had a lot of new stuff we hadn't done before, both hardware and software. Come the end of the month it was due, but we couldn't meet specifications on several issues. The customer was continually on our back reminding us of the huge sums it was going to cost him each day it was late. Sleep was lost. Luckily a couple of my bright young engineers came up with a solution I'd never have thought of. We shipped it out and got it accepted two weeks late. Incidentally, after forty plus years of my love/hate relations with all sorts of project engineering, that one pulled the chain on my retirement.

R. H. Eberstadt

My career has flown in part along a normal pattern, that is, a fairly high successive number of assignment changes, especially during the early years. All of this within a relatively small number of changes in corporate affiliations, only three during my career. The unusual aspect is what I think is an atypically high number of different superiors, and this generally without my changing assignments. During a career that basically spans forty-five years, I can recall between sixteen and different "bosses," resulting in an arithmetical average of between 2.2 - 2.8 years per "boss."

William H. J. Kitchen

Probably the greatest concern in being a "sole practitioner" is not having an organization to review the designs before it is too late to correct and/or change before the "extras" are presented to the "owners." There is another problem, too: making money is just as great an "art form" as engineering. If one wishes to make money, then it is necessary to enter into an arrangement with a "business minded" associate; however, a year or so later you can find yourself relieved of your command.

Harold Alden Wheeler

I enjoyed the privilege of working with the Wheeler Laboratory group of young engineers which were a carry-over from my company. As president of Wheeler Labs, I was in management but I was largely relieved of the problems of management by our contractors and my staff.

In general, I enjoyed much freedom and

remarkable appreciation of my work. I was mostly self-propelled, beyond the call of duty, because I enjoyed my work. It was a challenge to respond to my opportunities and to keep up with my ambitious and talented staff. I sometimes wonder if I should have provided them with more direct encouragement and freedom. In retrospect, I find myself more critical of my deficiencies than those of my superiors and subordinates. In Wheeler Laboratories, of course, my only "superiors" were the groups who were steering our work to meet their objectives. There my experience was very gratifying.

Before the war, we never had time cards. Later generations of engineers cannot imagine the freedom we used to enjoy in my company. It was not so in the larger companies, especially in the telephone company that was subject to regulation.

There were few occasions when my proposals were not welcomed, usually for sound reasons. My more vivid recollections relate to the few occasions when I rejected a sound proposal by a member of my staff. In such cases, I tended to be short-sighted, not appreciating the opportunities that should be expected with progress.

John Alrich

I much preferred California, so I looked in the want ads. I spotted a firm (still in Pasadena) that I knew something about, based on my earlier work for Bendix, Consolidated Electrodynamics Corporation (CEC). It was 1951 and this was where my career really got rolling.

CEC made two major products of fine quality and reputation; photographic stripchart recorders and mass spectrometers. Their technical staff was an excellent combination of talented engineers and physicists and was well supported with a good manufacturing facility; perhaps three hundred to four hundred people in all. Also, its president, Phil Fogg, was a man of great business courage and vision.

Fogg had been informed by several of his technical people that their method of data reduction for their mass spectrometers, using an analog computer of in-house design, was becoming time-consuming and of limited accuracy. The answer was to design something like the machine John von Neumann and some of his people were completing back at Princeton, New Jersey. Since back then there were probably less than several hundred people in the U.S. (perhaps more in Britain due to Turing and the Enigma cipher-breakers) who knew a great deal about this new technology, Fogg and his advisors probably underestimated the difficulty of what was proposed. At any rate, CEC decided to sponsor this development without much fanfare and began hiring for the program.

I was the second person hired for this aspiring computer group and there were about a half-dozen professionals from CEC who worked part or full time on the program initially. This included an excellent mathematician, a physicist, Cliff Berry, and a Program Manager, Martin Shuler, who had worked on radars during the war. They knew no more about digital computers than I did, which was exactly nothing.

Early on, CEC hired two part-time consultants who had computer experience; Dr. Harry Huskey who was developing the SWAC at UCLA for the National Bureau of Standards, and Dr. Ernst Selmer (from Norway) who had worked with John von Neumann and happened to be teaching for a short time at Cal Tech. Also as part of our education, there was one text that we knew about from which we could learn the elements of computer design (*High-Speed Computing Devices* by the staff of Engineering Research Associates, Inc., McGraw-Hill, 1950). So this book started out as our "bible."

Huskey gave evening seminars once a week but Selmer did virtually all of the logic design, now usually referred to as "architecture." The rest of us scrambled along as best we could developing the circuitry and modules.

It was sometime during this period that CEC spun us off as a wholly-owned subsidiary with our new name, "ElectroData." We were on our own, at least as a cost-center.

I became program manager of the arithmetic section which included everything except the drum memory, the paper tape reader and punch, the console, the magnetic tape drives, the typewriter control unit, punched card equipment and the power supply. I won't go into a description since this was published in the *IRE Transactions on Electronic Computers* (John Alrich, "Engineering Description of the ElectroData Computer," March 1955, vol. EC-4, No.1) except to say that the main-frame was largely composed of one hundred and seventy-three plug-in vacuum-tube modules, eight tubes per module (usually dual triodes), arranged in an air-cooled cabinet about twelve feet long, twenty-eight inches deep, and seventy-eight inches high. Quite an impressive sight when all the tubes were lit!

Its operating mode was as a single address, fixed point, binary-coded-decimal machine with numbers represented as absolute value and sign, shifted serially by half-byte. I mention this level of detail because of what happened.

Shortly after shipment of the computers (now called the Datatron 201 or the later model, the Datatron 205) began around 1954, usually to large corporations and universities, a significant advance in capability was requested (in at least one case, demanded!) by the scientific users. Burroughs, which now owned ElectroData, decided to comply partly because we thought other firms were also working on this feature.

Virtually all machines in those days operated fixed-point internally and most were binary machines. Floating Point (FP) operations were done by programming special subroutines for add, subtract, multiply, divide and conversion from fixed point to floating point and the reverse. In fact, using IBM punched cards and conventional key-punch equipment, the FP work format was already pretty well established and FP arithmetic was being done on electromechanical calculators. (These calculations were very time-consuming and, hence, used only where absolutely necessary.)

The field making up the word, in our case ten BCD digits plus sign, was divided into a two-digit exponent to the base ten and an eight digit mantissa, always less than one. Before each floating point addition or subtraction, the exponents of each operand were compared. If they were not equal, the mantissa of the smaller number was shifted right and its smaller exponent incremented until both exponents were equal.

In a similar manner, multiply and divide were implemented, taking care that exponentiation was proper and the mantissa was in formal form after completion of the operation. With this procedure, no change to the mainframe was needed except for the added circuitry. The range of a number was increased by fifty orders of magnitude with a penalty of two orders of magnitude in resolution. The speed of operation was improved considerably over the classical subroutine method, or course—probably by several orders of magnitude.

The scientific programmers, who were the heavy weights, were ecstatic over what this new capability meant in their work. What follows is generally true but my remembrance of some of the details may have suffered with time:

When a scientific sale was in the offing, usually a Ph.D. in mathematics was sent to the customer's site where they could discuss the problem one-on-one. These non-commercial customers were relatively rare and internal FP operations were non-existent so far as we knew at that time. Therefore, I had a completely free hand as to cost and performance subject only to very mild constraints, self imposed—the unit should be made of standard enclosure and modules already in use on the mainframe as much as possible. It should be capable of being retrofitted in the field, and it should satisfy our senior programmers in technique and performance.

It took about a year to build a working prototype. This was one of the last projects within Burroughs (or within any computer company for that matter) which was *not* designed by a committee! Our marketing people generally did not know what floating-point operations were since most of their customers were commercial users. The finished unit was styled exactly like the main-frame and bolted onto one end of the cabinet. It had about thirty-five plug-in modules, most of which were the same design as those in the CPU. There was enough "space" in our command structure to add the six new commands.

Cost was secondary since no other FPC was available as a standard unit when we started our design. IBM, Burroughs' great competitor, announced an add-on for its Model 650 shortly after my prototype was finished. I remember our marketing vice president, who had tentatively set the cost of the FPC at around twenty thousand dollars, immediately bumped it up to twenty-two thousand five hundred dollars after IBM announced their version would be available for twenty-five thousand dollars or thereabouts. I think we also shipped our first FPC before IBM did as well. I don't think many engineers today have as much fun as we did back in the late '50s.

W. Ross Appleman

I registered at the University of Illinois in the Fall of 1924 and received my BS in 1928 and my electrical engineering degree in 1936. So much of what has become common was new. A man came to Marathon Electric Mfg. Corporation where I was Chief Engineer and he needed a motor. I designed one for him and he became a good customer. When our salesmen in the Minneapolis office took me to a small machine shop owned by Mr. Jacuzzi, I had no idea I was designing a special motor for an invention with worldwide demand.

I also remember Mr. Bates, manufacturer of hospital beds, bringing a hospital bed to our Master Electric plant in Dayton, Ohio. He said,

"I want a motor to raise and lower this bed with a three hundred and fifty pound man lying in it. It must be very quiet. It must be very smooth, no jerking." This was so successful that later he brought a bed that was also able to raise the head and the feet independently. Today we accept these as standard equipment.

Paul D. Andrews

In 1923, the Post Office began flying some of their first class mail by plane. There were no radio navigation aids, not even radio communication with the ground, thus, the pilots had to fly low and maintain visual contact with the ground most of the time. In bad weather, this resulted in crashes at times. I was assigned the job of developing a radio transmitter and associated equipment which would at least enable pilots to maintain verbal contact with the ground.

I developed a small transmitter providing 10 watts of radio telephone output to a trailing wire antenna which the pilot could raise and lower from an insulated reel. High voltage for the transmitter tubes was furnished by a wind-driven generator, to be mounted out on the lower wing of the biplane. A control unit provided switching from the sending to the receiving functions, and for plugging in the microphone and the headphones.

An airmail plane was flown to Schenectady (NY). The equipment was installed and then test-flown by a well known pilot at that time. Everything worked fine, and I think that we sold quite a few to the Post Office Department.

Thomas M. Austin

In the Bureau of Reclamation, we usually worked in teams on projects which had been authorized by Congress. When I arrived at the Transmission Line Design Section, standards had been defined for each voltage. We could modify them slightly and develop better methods of implementing them. A suggestion system was in effect which encouraged individuals to make innovative suggestions, and rewarded the individual with money for the suggestions which were adopted.

I wrote papers on some of our innovations and designs. Some of these papers were published in the *AIEE/IEEE Transactions*. When a paper was accepted by the IEEE, I was sent, expenses paid, to deliver the paper at a meeting. In data processing, we had more freedom as computer use was new to the Bureau and we were developing methods of problem analysis and computer programs to solve the problems.

Sidney Bertram

In 1942, with the Antenna Laboratory off to a good start and WWII in progress, I joined the University of California Division of War Research at San Diego. There I was given the job of converting an experimental FM "fire-control" sonar to a "PPI Scanning Sonar." During the summer of 1945, nine submarines used the sonar to penetrate the mine fields. They also entered the Sea of Japan where they sunk many Japanese ships. They also aided the surface fleet by mapping other mine fields. I received a Bureau of Ships Citation for the invention of the scanning switch that made the outputs of the twenty range filters sequentially available to the PPI display.

Alan Bate

When the John Oster Manufacturing Company became the Oster Division of the Sunbeam Corporation, I was elected Vice President of Engineering. (Later I became a director of the company.)

I directed the designs for the cordless barber clipper, mixer, blender, food processor, can opener, slicer, shredder meat grinder, juicer, and the barium ferrite permanent magnets for series motors.

I helped organize the Association of Home Appliance Manufacturing's Member of Engineering Committee. I fostered professional engineering registration among R&E staff. I had five registered in an engineering department of one hundred twenty.

In immediate years before retirement, I testified in many product liability suits and testified in hearings before Wisconsin hearings on revision of tort law relative to product liability.

Yardley Beers

National Bureau of Standards, Division Chief, 1961-1968. When I joined the staff, the laboratories of the National Bureau of Standards (NBS) in Boulder, Colorado seemed almost like a university except that the experimental facilities were much better. There was a great deal of opportunity to do basic research of one's choosing, and there were opportunities to teach, both in collaborative projects with neighboring universities and in in-house courses. This picture perfect situation I now realize was the accident of economic conditions. Jobs for physicists were abundant. To attract staff in the face of statutory limitations on the salaries that could be paid, NBS had to offer other attractions.

Originally I was appointed as chief of a section doing research in millimeter waves, a field which captivated my imagination. Probably I would have been reasonably successful had I continued there. A few months after I started, in the first of the reorganizations that I was to encounter, I was promoted to my level of incompetence as chief of the newly formed Radio Standards Physics Division.

There were many superficial attractions to my position as division chief. The technical program of the division, which was of great interest to me, included the following items: research on and maintenance of the atomic clock; operation of the radio stations that broadcast standard time and frequency signals; research in materials at radio frequencies, plasma, lasers, and millimeter waves.

The division had about one hundred employees and an annual budget of two million dollars. I had a personal staff and a large office. People deferred to me as an important person. My immediate supervisor, John M. Richardson, sent me on several very educational trips, including a trip to an electronics conference in the USSR and a very meaningful conference for government science administrators held in Williamsburg, Virginia, by the Brookings Institute.

Soon I began to wonder just what my duties were in spite of an elaborate document which was supposed to define them. (Something

I did not have at NYU). No longer did I enjoy Saville's clearly defined delegation of authority. My decisions were frequently questioned, and some routine ones were countermanded by upper management.

Some of the problem was that my division and another, with closely related work, had an extra echelon of management between us and top management that other divisions did not have. This measure had been useful as a temporary one when we were created but should have been discontinued. My office devoted a great deal of effort in sending Washington information already in its files, only in a new format, often with capricious deadlines. The government was continually introducing restrictions which hampered us in doing our job. I can look back with satisfaction to only a few tasks I performed as a division chief, and the importance of my contributions to those tasks was unrecognized by my superiors.

I had always had clear plans for my own research, but I had no interest or proficiency in planning the programs of others. This was my level of incompetence. Not to lose job openings, I had approved the appointment of persons interested and qualified to do sophisticated jobs that were of low priority. They had done these tasks very well. In time I became accountable for our failure to accomplish higher priority work. Now it was now my duty either to forcibly reprogram the staff to work on priority tasks or to replace them by others who would as they became available. To the extent I was aware of this situation, I had no stomach for it.

In retrospect, my highest loyalty was to the staff, followed closely to my loyalty to my job description. My next loyalty was to my immediate superior, and my lowest loyalty was to higher superiors. The fact that at any time my superiors could give me a new job description I considered irrelevant and not the basis of personal loyalty. If their actions hindered my carrying out my present job, I opposed them.

In an acrimonious telephone conversation with someone in Washington management, I stated my objections to some measure he was instituting. I said that, since it was impeding our work, I would have to take a specified countermeasure.

He responded, "Do you mean to say you are threatening the President of the United States?" I was shocked. Later, I realized the logic of his question. I was an officer of the Executive Branch of the government, which was carrying out the provisions of the Constitution. I had received my authority by a series of delegations from the President. By threatening someone higher in the chain, in principle, I was threatening the President.

This reasoning leads me to wonder if anyone who believes in the laws of nature ever can give his or her highest loyalty to any government, and whether there is a proper place for truly basic research in any government organization. I did not have to actively face this moral question for long because shortly I was relieved of my position as division chief. I was given a job as a consultant, which was the type of job I wanted when I first entered the Bureau.

William S. Cranmer

This type of combined business (accounting degree) and technical background (MS in engineering management) is only mentioned as it had a pronounced effect on my engineering career. In the early days of the electron tube industry, I was totally impressed by the genius of the more senior engineering personnel. It was commonplace to see engineers walk down a hallway with vacant stares as they concentrated on some function or equation, completely unaware of their surroundings.

On one occasion, I remember approaching one of these individuals and asking a legitimate technical question. The engineer stopped, concentrated or a moment, and then just walked away. I thought I had interrupted him at a wrong moment and went back to my place of work.

Three and a half weeks later, this engineer came up to me and with no introductory remarks gave me the answer to my technical question. Many of these geniuses would do their best work with their feet propped up on top of their desks!

During those years, I believe I completed some worthwhile engineering tasks and received only occasional rebuffs from my superiors. One I do remember, however, was a technical presentation given by an accomplished engineering leader. After his presentation, the speaker approached me and asked, "Did you understand that lecture, Bill?" I replied that I did understand a good portion of it and congratulated him on a well organized presentation. He said, and I quote, "Well I thought I would ask you, Bill, because if you understood it, anyone would understand it!" and he walked away. For the most part, I am happy to report that I was fully accepted as a part of the engineering staff.

One of the greatest educational benefits was the opportunity to represent my company at JEDEC, IEEE or Armed Services product meetings. It took a bit of training to gain experience and self confidence at such meetings but the rewards were substantial. At one of the Armed Services meetings I attended as a new member, the Chairman announced, "Don't worry gentlemen, in a minute we are going to hear from RCA." I would look around the room and suddenly notice that I was the only one there representing my company and they meant me!

It was a challenge, but over time I managed to learn the language and the proper protocol for these occasions. After numerous meetings, I was finally able to speak up and say such things as, "Mr. Chairman, I request that the secretary include in the minutes a report of my minority opinion on this last point in question."

Other meetings were in-house "strategy" meetings to insure that everything was in order in preparation for an important forthcoming meeting with a customer. I recall one marketing manager saying, "When you engineers get to the meeting just sit there and don't say anything or you will blow the whole contract---." I am proud to say that we engineers talked anyway and came up with some valuable comments for all concerned.

Edgar C. Gentle, Jr.

Competition? There was mostly friendly competition, but some of it was not too friendly! But organizations do have a lot of competition. I was rewarded I guess for doing jobs well. I was fortunate.

How important was the personality of myself, my peers and of my team supervisors? They were and are naturally very important. In large organizations, I learned very early, it does require give and take and it's equally important to work compatibly with various personalities.

Tact is essential in my opinion. You don't have to yield to personalities with whom you may conflict, at least I found you don't. One way to get around personality clashes is not to discuss on a personal basis those things on which personalities clash. This is easier said than done—many items can be important to the job on which you are working; but endless discussions really accomplish very little when there is a personality clash. I think you have to guard against these types of discussions.

Attitudes of people are very influential and they can help a person's work. These attitudes can also hinder work and it's just something I believe that an engineer or whomever else as an individual has to face up to; an individual has to find a way to talk to peers and try to sell ideas. If an individual just cannot, then I presume that person will move on to something else and let someone else deal with and handle it.

In retrospect, I expect my greatest accomplishment was the contribution I may have been able to make in improving telecommunications in the Bell System across the United States. For example, adopting new methodologies to try to improve productivity in operations and to keep service at high standards, to provide additional employee opportunities and to reduce costs. Another accomplishment was the ten years during which I was responsible for the revenue requirements and regulatory activities of South Central Bell. I was able to put together explanations that could be effectively understood by the State Public Service Commissions to make sure that customers, employees and owners of South Central Bell (with AT&T as the surrogate for South Central Bell) were treated fairly.

Worse failure? One situation, which fits that description, followed an extremely heavy rainfall. A major city was totally isolated from the Public Switched Network. That's the largest failure with which I have ever been personally associated and it was in my department so, of course, I was responsible.

The least encouraging sometimes are peers. As to what motivates people to discourage others, I would be afraid to make a guess.

John F. Bell

After WWII ended, war contracts also ended abruptly. And a conversion back to civilian production had to be made immediately. Small radios were put into production right away, but the real emphasis was on resuming development of television standards and television receivers while full scale radio production was building up. Pre-war black and white television standards and receiver design were well underway, but were seriously deficient regarding immunity to interference.

TV transmissions began and receivers soon were being manufactured and sold. Interim standards were finalized and soon very high frequency (VHF) receivers were designed and being sold in large quantities. However, seventy ultra high frequency (UHF) channels were also assigned and color television was being developed. These remained to be incorporated into the standards. The additional UHF channels were needed to support a nation-wide TV service.

The first TV receivers manufactured were

designed like the early experimental sets with a twenty megahertz intermediate frequency and inadequate preselection and shielding. Excessive oscillator radiation, direct IF (Intermediate Frequency) pick-up, and image interference was therefore inevitable. When these receivers were sold in sufficient quantity to heavily populated neighborhoods, severe interference made it evident that the receivers had to be redesigned, and the channel assignments had to be coordinated with the revised receiver design to avoid such interference.

The early receivers were only for VHF. The public had to buy new receivers or the almost impractical UHF converters to receive the new UHF channels. Since there were large numbers of VHF only receivers in the field, a way to receive UHF on these and the redesigned receivers was badly needed.

New receiver designs were in progress incorporating the new forty megahertz intermediate frequency, and also improved preselection and shielding. Early on I had announced that, using some very compact UHF resonant circuits I had developed, it was possible to build a single channel UHF strip that would directly replace the VHF channel strips. This would enable the existing receivers to receive UHF channels and operate on UHF in exactly the same way as they did on VHF. Since the VHF vacuum tube amplifiers would not work on UHF, the UHF circuits and the mixer had to be on the channel strip. The vacuum tube RF amplifier and mixer had to operate on the intermediate frequency to provide the needed additional amplification following the crystal mixer.

After we had designed such a strip, we were in a position to equip all sets in the field to receive the new UHF channels then coming on the air. As each new model TV receiver was developed, a new UHF channel strip was designed for it. Subsequent models were equipped with a UHF continuous tuner. Sometimes UHF strips were used on sets equipped with an all-channel tuner to maintain the turret tuner convenience on frequently used UHF channels.

Because of the serious field problems from oscillator radiation and the image and direct IF pickup interference, the FCC ordered all receiver production stopped. The FCC also stopped licensing new transmitters—the famous three year "freeze." Some important markets were left with no TV service, among them Denver, Colorado and Portland, Oregon. This resulted in many complaints and letters to Representatives and Senators. RCA moved its experimental low power New Haven (CT), UHF transmitter to Portland to provide interim service.

Many important system parameters remained to be settled, and the freeze lasted much longer than had been generally anticipated. This was because the job of setting up a sufficiently coordinated set of standards was much more involved and controversial than expected. There were competing color systems and all systems had to be tested and a choice made. Also, the whole system had to function in the six megahertz channel already in use which had not anticipated color.

Squeezing additional chrominance information into the existing channel seemed impossible. However, the eye cannot perceive fine chrominance detail. Thus, a practical solution evolved that added the color information into the six megahertz channel and still met the compatibility requirements. Monochrome receivers would produce a full definition monochrome picture on a color broadcast, a color receiver would produce a full definition monochrome picture on a monochrome broadcast. What's more, the additional color information modulated onto the carrier within the six megahertz channel.

Therefore, the picture information was divided into two components: a luminance component which alone produced a full definition monochrome picture, and two low definition chrominance components. These added to the luminance and produced what appeared to the eye to be a full definition color picture.

Kenneth Sturley

To my delight, the Marconi Company offered me a research post. After an introductory course under A. W. Ladner (author of *Short-Wave Wireless Communication*) at the Marconi College, I began the design of a rebroadcast receiver with directional aerial pick-up. Just as this was completed, the 1930s slump hit Marconi. The forty youngest researchers had to go. Each of us was interviewed by Research Head, G. M. Right (yes, father of "The Spycatcher"), and given three months advance salary, typical of the family atmosphere at Marconi.

Four years later, a surprise letter from Ladner invited me to return as a lecturer at Marconi College. (The college specialized in training the engineers of customers for operating all kinds of Marconi communication equipment.) I was able to pursue work on receiver circuit analysis and publish my book, *Radio Receiver Design*. The college made an important contribution to radio communication training in World War II.

With the war's end came Ladner's retirement and an offer of the College Principalship, but an invitation from the British Broadcasting Corporation (BBC) to start a technical training department proved more attractive. The next seventeen years were very rewarding under the light rein and staunch support of Roland Wynn, BBC's Chief Engineer.

Life was pleasant at Wood Norton (near Stratford on Avon), the former chateau-style home of the French Pretender, Duc d'Orleans, turned into a wartime standby and fully equipped for sound broadcasting. My wife aptly described my situation as "Lord of the manor with none of his financial embarrassments." Equipment for black and white, and color TV training was added later and courses were organized for trainee technicians, promotee technicians, graduate engineers and for up-dating senior staff. Students came from Commonwealth countries, and there was a succession of visitors from Europe and overseas who wished to start broadcasting training schemes.

In the 1950s, I visited the United States to examine the electrical engineering training. I was impressed by the correspondence courses and universities such as MIT. My hopes of a sabbatical term at a U.S. university were not fulfilled; but F. J. Terman, in his reply, boosted my ego by adding, "I have read your book."

K. L. Rao

The year 1959 saw the introduction of TV in Delhi. By the 70s, TV stations were set up at Bombay, Calcutta and Madras. At this stage, a new region—southern region—was created to look after the development of the services in the Southern Region. The work involved the development, installation and maintenance of sound and TV broadcasting stations in the south.

In the sixties, I went over to the research department. Lack of low-cost receivers—within reach of the common man—was inhibiting the rapid growth of broadcasting in India. Hence, I was directly involved in the development of low-cost receivers suited to our environment. During the course of time, I was involved in various fields relating to the development of broadcasting in India.

I was posted as Deputy Regional Engineer at Madras which is the headquarters of this region. I was there in a supervisory capacity. In this capacity, I retired in 1972 after thirty-one years of service. I might, perhaps, have been better off, but I have no regrets. My interest in broadcasting was so great that I was able to rough out several difficult situations. I have had the satisfaction of having done my duty well.

Franklin Offner

My own research was what led me to the field: to find out the physical basis of nerve conduction. It was known that cell membranes, and therefore presumably nerve fibers, had both resistance and capacitance. Also, that the membranes were semi-permeable to K^+ ions resulting in a negative internal potential.

I had the hypothesis that nerve conduction resulted from the development of a voltage-sensitive "leakage" conductance across the membrane. I solved the partial differential equation for such a system, and showed that it would in fact result in a propagated impulse. The predicted impedance change was found by Cole and Curtis in squid axons; fifteen years later, Hodgkin and Huxley made a complete experimental and theoretical analysis of the phenomenon, with my model as their starting point, work for which they received the Nobel Prize.

In my Ph.D. work, I was sponsored by the great physicist, Carl Eckert, who gave me full support and counsel. He was a man of probably the broadest intellect of anyone I have ever known, and also the most helpful and unassuming. After receiving my Ph.D. in 1938, I went to New York City for nine months to assist a company which had taken a license on my EKG (electrocardiograph) patent.

I then returned to Chicago, and formally started my company, Offner Electronics, intending to primarily manufacture electrophysiological equipment, such as the EEG (electroencephalograph); in fact, I had received a request from Denmark for three such instruments even before I had decided to start the business, for which I had five hundred dollars capital.

Besides the EEG, I had been asked by a psychiatrist to build an apparatus for giving electroshock treatments. Although I was very dubious about the value—and the safety—of such therapy, I consented to develop such an apparatus, which was then not available in the U.S. I developed an apparatus which permitted the physician to provide the treatment with maximum safety to the patient. This apparatus became widely accepted in the Americas and Europe; and contrary to my initial belief, proved to be a highly effective treatment.

With the fall of France, I turned my efforts to defense, essentially to the exclusion of all else. In the summer of 1940, I proposed to the Navy (through the NDRC) a method of locating submarines by dropping buoys with sonic pick-up transducers, which would transmit the signal to the over-flying aircraft. I received a reply stating that my suggestion was very important, and that I should disclose it to no one else. This was the origin of the Sonabuoy, still our primary antisubmarine weapon.

Concurrently, the Heat Research Laboratory of MIT was attempting to design a heat-homing high angle bomb. But the lab was unable to develop a heat seeker having either sufficient reliability or sensitivity. The Air Force suggested that I attack the problem. I redesigned their seeker, again using the technique of cross-correlation. In two weeks, I produced a seeker that was far simpler, reliable, and, according to the tests of the Heat Research Laboratory, at least ten times more sensitive than their design. Their bomb was then put into production. After the war, the Sidewinder IR guided air-to-air missile used, and continues to use, the same principle in its seeker. My company also developed and produced most of the Geiger counter electronics used in the Manhattan project, as well as a number of other developments for the Air Force and Navy.

When experimental Ge (germanium) junction transistors first became available in 1952, they were said to have many problems. Nonetheless, I purchased three transistors from RCA (they were tightly rationed), and started developing circuits, based on *their* characteristics, not on vacuum tube technology. Within a few days, I had concluded that vacuum tubes would soon be obsolescent. Accordingly, I told the engineers in my company that we would stop all development work on new vacuum tube-based instruments, and start a program on transistorization of our full product line.

The first instrument we attacked was probably the most difficult we could have chosen: the EEG, because of the high sensitivity required (down to 1 mV), with a frequency response down to 1 Hz or less. The completed instrument, designated the "Type T" EEG, was portable, with eight recording channels. The Type T revolutionized EEG, not so much because of its portability, but because of its unprecedented reliability. It won almost immediate acceptance around the world; the British manufacturer of EEG's left the field, being unable to compete. In 1961, Offner Electronics Incorporated merged with Beckman Instruments, which I left in 1963 to return to academia.

Robert Newman

This personal reminiscence may sound like an old soldier telling his son how he won the war single-handedly. The vaguely recalled past often appears that way.

Our TV tube business (General Electric) suffered from a common disease: we wanted to be totally different from everyone else. Thus our first tube was a high voltage Schmidt-optics tube—part of a 1946 large screen TV receiver retailing at one thousand dollars. It received little support from dealers and ultimately from customers—we were much too early.

So we argued that small TV sets would become a hot product: one 7.5 inch set for each family member. Here again we were out of step: a decade too early. The Buffalo shop was filled with unsold inventory: large screens and small screens but no saleable screen size tubes. (Note: The tube had a metal cone, and was difficult to manufacture. We might well have lost money even if sales were high.)

The Korean adventure found us making many electronic devices for the military, including a small microwave tube resembling a lighthouse, and not unnaturally called a Lighthouse Tube. It was the central item in the IFF system—Identity Friend or Foe. This caused a plane's guns to shoot down an approaching plane if it did not give the correct coded response to its automatic inquiry. We were the only supplier of the tubes.

Early one Monday morning, the engineer I had charged with surveillance over this product line came to my office breathless, "The *LIFE TESTS* in the Lighthouse line were all failing." Sample tubes during testing did not meet specifications after being operated for a period.

This meant the previous three weeks production had to be scrapped, and that something—some unknown process—was out of control. We daily fended off Dr. Baker, assistant secretaries of the Navy and endless experts from the Research Lab and other Company areas while we tried to solve the problem. Neither my staff or I were getting more than a couple of hours sleep each night. A major difficulty was that it took nearly two weeks before we knew whether a set of experimental measures we were taking had solved the problem.

Finally, just as our automated planes were destined to destroy each other, I got a clue: a small mica spacer had a contamination which was not removed by our usual 200 C air drying and had to be heated to 400 C. We began the new process. I dashed off to visit the mica part vendor, not being satisfied with his phone comment that nothing in producing the part had changed.

Walking through the shop I noticed a punch press operator—where the mica was being punched—brushing the die with material from a can with a GE label. "Silicone oil," said the operator, "your silicone salesman said this would improve die life, and cause less contamination of the parts."

So we were nearly defeated (both our department and the U.S. Navy) by a brother department whose technical sales people ought to have recognized that our vacuum tube operation was *set-up* to handle the easy to decompose oils formerly used; silicones didn't decompose as easily. In fact, the silicones actually slowly poisoned the finished tubes.

The only response of my department General Manager, George Henyon was, "Bob, why didn't you think to remove the silicone earlier?"

Robert McLane

The total experience at Honeywell was one of immense variety, growth, responsibility and challenge. I noted many engineers who were promoted into management positions and were very happy and successful. I also saw some, who were misfits and had no "knack" of management, fail badly because they didn't really enjoy the change.

Honeywell was in a rapid growth phase in these years, and soon was giving specific attention to the needs of training engineers to become good managers. My career had peaked out at the supervisory level when I was transferred from research engineering to systems engineering into a position promised to another. Both the "promisor" and "promisee" soon left the company, and I started anew in my specialty!

This specialty evolved from skills acquired in the Research Department in automatic beam guidance and blind landing systems for all-weather routine safe operations. It culminated in team participation in U.S. Patents 3,015,459; 3,040,568; 3,045,955; and sole patent 3,055,214...all in 1962. The close work with test pilots in these programs moved my specialty into man-in-the-loop human factors research and engineering. Providing engineering environmental simulations for human performance evaluations by Ph.D. psychologists was a most broadening experience, leading to 1966 National Symposium Chair, IEEE Human Factors in Electronics, and Vice-Chair IEEE Group on Man-Machine Systems 1969-1970.

The Research Department environment was superb, due to the excellent directorship of one O. Hugo Schuck, followed by M. A. Sutton, and through W. T. Sacett, D. L. Markusen, R. N. Bretoi, and H. G. Sperling, as specific levels of management over the years. The team spirit under these leaders was outstanding, and led to individual development and responsibility and a real feeling of contribution to the overall successes achieved. The attitudes of the marketing arms (home office as well as out-of-state offices at centers of potential contract support) were excellent. So was the support of the patent attorneys, who closely monitored our memoranda and talked periodically with us to determine significant achievements worthy of application to the U.S. Patent Office. The engineering department's rapport developed by research management led to many pilot feasibility programs to determine practicality of promising ideas for productivity as market offerings.

Simpson Linke

During the sixties and seventies, my duties were equally divided between teaching and research. Dr. Henry G. Booker, director of the electrical engineering school in the early sixties, encouraged me to study arc phenomena in vacuum through the medium of the vacuum circuit breaker. For several years, R. N. Sudan and I had joint grants from the National Science Foundation (NSF) for these studies.

Also, during this period, a number of graduate projects in various energy systems were conducted under my direction. Thermionic converters, thermoelectric devices, hydrogen-oxygen fuel cells, an upgraded version of the Brookhaven Laboratory mercury pump, and an axial-air-gap induction motor with a solid-disk rotor were designed, constructed, and tested with varying degrees of success.

With the able assistance of E. C. Ogbuobiri, one of my graduate students who was an early expert in computer techniques, a load-flow algorithm was developed for use as a laboratory tool in my power-system courses. Unlike the somewhat remote contact afforded by the old ac calculator, this new approach allowed students to have direct interactions with their system analyses which resulted in enthusiastic appreciation of the computer approach to power-system studies.

Starting in 1981, I served a three-and-one-half -year term as Coordinator of Graduate

Studies in Electrical Engineering. In this period, I also assisted in the establishment of the electrical engineering school's Kettering Power System Laboratory.

In accordance with the general adoption of new computer facilities in the school, I removed my power-system-analysis algorithms from the university main-frame computer and installed them on the Kettering VAX computer. This step changed off-line card programming to on-line remote-terminal access and resulted in a dramatic improvement in students' analytical capabilities and interests in the solution of power-system problems.

Paul Burk

I installed the first FM police radio in Juarez about 1950, and about a year later a similar system on the same frequency in Chuahuahua City. Shortly after this installation, a gang of bank robbers in Juarez was making a high speed getaway heading south, ahead of the Juarez police who had no hope of catching them. So the Juarez dispatcher telephoned the Chuahuahua City police to send several radio patrol cars north to intercept the gang. About a half an hour out of the city, the two police forces were in contact with each other and was that gang surprised at the "friendly" greeting they got going and coming! The Juarez police chief felt that one job paid for the whole radio system.

Leslie Balter

One of my assignments in 1941 was to quiet the first data processing installations assigned to the First Army. State of the art automation in 1941 meant using the IBM punch card. The first military use that I knew of was getting updated strength reports for field units. The punch card was used to track field casualties and provide updated strength reports.

They ran into a problem, however, since the data processing systems at that time consisted of the 080 Sorter, 024 Keypunch, 077 Collator, 602A Calculator, 402 Accounting Machine, and so forth. All programmed via wired panels and all using electrical contact impulses to read the punched holes. This produced tremendous electrical interference which made radio communications virtually impossible. Since the data processing equipment would always be located at the Army's command center, where of course the communication equipment was located, they had a "big" problem.

In 1942, I was head of a team sent to Endicott, New York to work on MRU #1 (Machine Record Unit #1) assigned to the First Army. Our objective was to suppress the radio interference produced by the EAM (Electronic Accounting Machines) as they were known. We spent about two weeks at Endicott and had the unique experience of residing in the IBM Guest House on the golf course. I won't use the word pleasure, because visiting and being a guest during the reign of Tom Watson, Sr., really was an experience. Guest or not, the doors were locked at ten p.m.—no liquor, and proper decorum at all times.

In that period, every IBMer wore a gray flannel suit, a white shirt, a grey felt hat and towed the mark. They were supposed to do as the signs all over the plant stated, "THINK." And there were "THINK" signs everywhere. No IBMer, in good standing, would be seen drinking in a public bar. Papa Watson ran a "tight ship" and apparently it did good things for the company. Our group did its job and MRU #l went on to First Army Headquarters and did what it was intended to do.

At the war's end, there were virtually no engineering jobs available. The war plants were being shut down and engineers were looking for new fields to conquer.

The big new field was television. I managed to get a job in a private school teaching television construction, service and repair. There were millions of veterans returning home, looking for work and in need of new skills. Schools were springing up all over to satisfy this demand.

After teaching TV for about one year, I opened my own school in Jersey City, the "Jersey City Technical Institute," licensed in May, 1947. We taught television. We taught theory and provided lab experience including building a TV from components. The students learned theory, construction and then servicing to get the TV operating properly.

Since we taught electronic wiring, applicants wanted to learn "IBM" wiring. In 1958, I opened the first "IBM School" in the state of New Jersey. We taught control panel wiring for the early electronic accounting machines (602A Calculator, 402 Accounting Machine, 077 Collator, etc.). My order with IBM was the first for training purposes.

Herbert Butler

In 1950, work was done for the Air Force under a RAND Corporation study dealing specifically with the concept of a television-equipped meteorological satellite. RAND then contracted with RCA to investigate how well a TV camera could operate in a space environment. The work involved several aspects of reconnaissance and surveillance, and fitted in very nicely with my work in Cyclops.

At the Army Ballistics Missile Agency in Huntsville, Alabama, Dr. Wernher von Braun became interested in RCA's work—specifically, in a proposal for a TV satellite submitted by RCA. I attended some of these conferences chaired by von Braun and returned to Fort Monmouth full of enthusiasm. I urged management to become more deeply involved since, traditionally, the Signal Corps had responsibility for weather service to the Army. It was not until ARPA (Advanced Research Projects Agency) was established in February of 1958 that the program became formalized as the Television Infra Red Observation Satellite—TIROS. The payload responsibility was assigned to the Signal Corps Engineering Laboratories, with responsibility for the launch vehicle to remain with ABMA.

An ad hoc committee on meteorology was created by ARPA in May to help formulate the design objectives for the satellite. I was assigned as the Signal Corps representative on the committee.

I became the Project Manager for the Signal Corps, responsible for the technical direction of the development of TIROS to be built under contract with RCA. This involved not only close coordination with the committee because the design of the system would go through considerable evolution, but also very close coordination with the Weather Bureau. As the prime user of the data output, the Weather Bureau was the "customer" who had to be satisfied. Military organizations were involved, not only because of interest in the data, but because of practical and security reasons requiring that the ground stations be installed at Air Force, Army and Navy locations.

The Infra-Red subsystem of TIROS was designed and engineered at the Signal Corps Labs by a team led by William G. Stroud, Dr. Rudolf Stampfl and Dr. Rudolf Hanel. Sometime after NASA was activated in October, 1958, most of the team transferred to the Goddard Space Flight Center, in Greenbelt, MD. Dr. Stampfl persuaded me to join them at Goddard.

TIROS I was launched and operated successfully on April 1, 1960. It marked a high point in my career and probably was my greatest accomplishment. The Department of the Army awarded me the Commendation of Meritorious Civil Service.

With the activation of NASA, program responsibility for TIROS was transferred to the new organization, but I continued to be heavily involved. There were many complexities to be resolved including: administrative, legal, contractual, funding, budgeting and scheduling. The downstream satellites had to be built, operation of the ground stations and the communications links all had to be maintained. Fortunately, I had good rapport with the people at Goddard. I worked quite closely with their new Project Manager, Robert M. Rados. Gradually, the new organization was able to assume control and I began to disengage, albeit somewhat reluctantly.

Wieslaw Barwicz

My first job in 1938 was in a Philips factory in Warsaw as the manager of a small department producing small Pupin coils for telecommunication purposes. At this time, Philips was looking for young Polish engineers and I decided to take it. My wife was working in Warsaw. This first job was very good for me because I started with Dutch machines waiting for set up. I started without training (probably no time). My responsibility as a manager of this small department was to set up machines, to teach some workers and make Pupin coils from special flattened iron and wind them.

I think that responsibility was just right for me. However, I did not choose this specialty especially since this activity had no future. In 1939, WWII began and I was taken to another department namely the Electron Tube Department. In this department, I worked as an engineer until the insurrection in Warsaw. In this department there were only two engineers. One for electrical measuring and me, principally for technology. The leader of this department was German and practically always absent. Therefore, I had a freehand. The Dutch chief manager wanted me to work from the beginning on all machines step by step. It was, of course, very good for my future.

Responsibility (Germans and war) was very high. For instance, some workers were taken to the unpleasant camp. Therefore, we had some troubles and friction. But it was the time of war and occupation. The relationship with the German leader was not pleasant.

When the war was finished in 1945 and Berlin was conquered, a lady from the Warsaw Ministry came to me with a proposal. She wanted me to undertake a terribly difficult task, namely to organize a Polish factory of electron tubes in the south of Poland because the Philips factory had been destroyed.

Realization of this proposition was at this time very, very difficult. The country was completely ruined—no machines, no technical documentation, no material, no specialists. It is necessary to underline that at this time I was the only engineer in Poland who knew about or who was a specialist in the technology of electron tubes (such a situation may be curious for western countries but it was true). The specialists from Philips factory were dispersed over the whole world.

The decision was very difficult for me to make. After deep meditation, I said all right. I went to a small town in the south of Poland where I found only two empty buildings. They had been the German Telefunken factory of electron tubes during the war. The production machines had been taken to Germany by German soldiers and the rest had been taken by Russian soldiers to Russia.

In this case, I had good luck because the Russian soldiers did not have enough room on their trucks so they threw some of the machines into ditches. These machines, after being repaired (which was very complicated), were used in the future factory. We also had good luck because the Russian authorities did not find the small laboratory situated in the south of Poland in the mountains.

This was very important because in this laboratory we found a power electron tube (the only one) needed for induction heating the electron tubes and for pumping automation. This laboratory had been destined for new elaborate metal ceramic tubes for military purposes. We also found a small quantity of raw materials. But the grid winding machines and automated pumping machines were missing. My expertise was electrical engineering not mechanical. What to do? I constructed both machines, and after building them, they were quite good. Especially the grid winding machines, they were still working many years later in the Warsaw factory.

For raw materials, we had to look out in the country. The most important was finding getter (for maintaining a vacuum in the pumped out tubes). I had luck because in in the ruins of Warsaw I found two small boxes of getter.

Now the question of specialists. Because in ruined Warsaw there were not enough houses, I invited some workers from the former Philips factory to come to the new start-up factory. About ten workers came including one chemist and two from the universities. Also some people who were already there joined. The latter would be trained.

With these "assets," I started step by step, day by day with the production of electron tubes. In the first two or three months, I was pumping tubes myself and performing many other activities. The first tube in our factory was of course the rectifying tube AZL. The next year, it was the power pentod AL4. After two years of production and learning, the factory was completely transported to Warsaw at the destroyed Philips factory. In Warsaw, we started with the production of old tube types and new tube types. Some years later, around a thousand workers worked in the factory.

Glydus Gregory

To be an engineer was never my ambition. I became an engineer after many years of work as a technician. Some of my bosses questioned whether I ever became an engineer. I held the title of "engineer" for about ten years before my retirement. However, I think I did as much engineering as a technician. I did get a little more into systems and management as an engineer.

My greatest help and my greatest hindrance came from the same person, Arnold H. Carver. He was my first, second or third-level supervisor during most of the time that I worked for the state of California. He was not an outstanding engineer, but as an organizer and expeditor, I never met his equal. His failing was that he would never follow the rule book if he could find a short cut. Also, he sometimes antagonized people when it was unnecessary or made promises that he could not keep. I tried to copy what I thought were his good points and to avoid his mistakes. I was later called on to expedite jobs which I think I did without antagonizing the people concerned.

Joseph E. Guidry

By a letter dated May 20, 1961, President John F. Kennedy requested the Secretary of the Interior, Stewart Udall, to review the International Joint Commission's Report of April, 1961 on the International Passamaquoddy Tidal Power Project and the advisability of hydroelectric power development on the Upper St. John River. The Secretary then appointed the "Passamaquoddy—St. John River Study Committee," chaired by Mr. Morgan D. Dubrow, Assistant and Chief Engineering Advisor to Mr. Kenneth Holum, Assistant Secretary of the Interior, Water and Power Development, and composed of experts from the various bureaus.

The committee appointed Joesph E. Guidry (myself), Assistant Chief, Programs and Development Branch, Power Division, and Bruce Rogers, Electrical Engineer, Bonneville Power Administration, to conduct a power load and resource study of the New England area of the United States and the Maritime Province area of the Dominion of Canada.

The load and resources study was made as requested. A report was made to the Chairman, Passamaqoddy—St. John Study Committee, December 13, 1961, stating the following:

"We are firmly convinced that Passamaquoddy should be developed to provide peaking power in the magnitude of one million kilo-watts and its operation should be coordinated with the full development of the St. John River hydroelectric system, including upstream storage in the state of Maine."

The committee accepted this conviction and proceeded in close collaboration with the U.S. Army Corps of Engineers while utilizing the technical expertise of the Bureau of Reclamation. Computer studies, design studies, engineering and economic studies were made. The results were included in a report from Secretary Stewart L. Udall to President Kennedy dated July 1, 1963. The President accepted this report on July 16, 1963, and requested the Department of the Interior and the U.S. Army Corps of Engineers to complete additional sensitive studies so that the report could be transmitted to Congress for authorization. The State Department was also directed to proceed with negotiations leading to agreements with Canada for the construction of this international project.

The Army-Interior Advisory Board was formed on July 30, 1963, and later augmented to include representatives from the Department of Commerce and the Federal Power Commission as participants, and representatives of the Bureau of the Budget, the President's Office of Science and Technology, the President's Office of Economic Affairs, and the Atomic Energy Commission, as observers. The Board contributed immeasurable in both the evaluation and investigation of the project and in perfecting the report.

On August 3, 1964, Morgan D. Dubrow, Chairman, and Joseph E. Guidry, Project Engineer, Passamaquoddy-Saint John River Study Committee, submitted to the Secretary of the Interior the results of this joint effort in the report, *Supplement to July 1963 Report The International Passamaquoddy TIDAL POWER PROJECT and UPPER ST JOHN RIVER hydroelectric Power Development, August 1964.*

Then Senate Bill, S. 2573, was introduced to authorize this project and Senator Edmund S. Muskie presided over a Hearing, 88th Congress, 2nd Session, US Senate, Subcommittee on Public Works, August 12, 1964, Washington, DC. The Passamaquoddy reports are reproduced in full in the *Committee Prints of the Hearing*.

On November 22, 1963, we were returning from an important Passamaquoddy meeting in Boston in a corps of engineer's plane when someone turned on the radio and we heard, "The President is receiving human blood," and then shortly thereafter, "The President is dead." We were speechless and shocked.

Although Senator Muskie and others strongly supported Passamaquoddy, the President's assassination was the death knoll for this significant constantly renewable energy resource.

The following quotes from Senator Edmund S. Muskie's letter to me on August 12, 1963, furnishes an insight into his feelings about the project, "Dear Joe: ...thank you...for the major contribution you have made to the successful Passamaquoddy-St. John Study Report. I am quite confident as to the outcome of this project. Much of the credit will go to you. I look forward to working with you in the future.....Sincerely, (signed Ed)."

Clive M. Gardam

My first position was with the English Electric Company of Canada, in St. Catharines, Ontario, a manufacturer of transformers, motors, and switchgears. An interesting project I was given during my two years in the Motors Department was to develop the design for a 25/60 Hz motor for the Hydro Electric Power Company (HEPC) of Ontario (now Ontario Hydro) to be used during the frequency change over from 25 to 60Hz. English Electric manufactured many of these motors for the HEPC.

Later positions with English Electric included two years in Switchgear Design and two years as Head of Quality Control; a very demanding job, as the company had been bought out by new owners, and morale was very low in the offices and in the shop. This, in effect, lead to my voluntary departure. I left in 1954, moving to the Toledo Edison Company where I worked until my retirement in 1984.

At the Toledo Edison Company, I received several promotions and held several positions from Assistant Engineer in Power Plant

Electrical Design, through Electrical Relaying and Control Engineer, System Planning Engineer, Substations Design Engineer, Manager of Station Electrical Engineering and finally to Director of Transmission and Substations Engineering in 1974.

Robert G. Johnson

My career began with formal training as a physicist with both a bachelor's degree and a Ph.D. (1952) in physics. After three years as project engineer developing microwave tubes, I joined Honeywell and was active in corporate research for the remainder of my career.

Among the accomplishments that stand out in retrospect, the development of reliable gas discharge ultraviolet photon detector tubes for flame detection and monitoring systems was one of the most interesting and significant. That activity combined gas discharge and electron emission physics with the engineering requirements of reliabilty and producibility, and resulted in an extensive and economically significant family of products.

Over the last three or four decades, I have seen the philosophy of corporate research change quite a bit from one of emphasis on basic studies and pioneering to one of much more direct responsiveness to expressed corporate needs. In effect, I believe that this is a shift from a long term view of the corporate future to a much shorter view. This has become sharply evident in recent years with the corporate raider leveraged buyout trend, which I view as a major disaster for stability and productive long term planning.

The earlier development of the photon detector tube flame sensor systems occurred in the early 1960s and was done with the full blessing of top management. On the other hand, I began the silicon chip mass airflow sensor research development about 1979, and it was vigorously opposed by management then as being a "flaky" idea. The opposition was not only a denial of funds so that development began on a bootleg basis, it also took the form of bad performance reviews and denial of raises.

This technology, and its derived devices, however, now accounts for about a quarter of the corporate research group's total research and development budget. The work is now fully recognized as a major activity with a high degree of relevance to the company's business future. The competitition in this technology throughout industry is intensifying. It is worthy to note that had I been a younger man when the "flaky" idea was opposed, I would doubtless have migrated to another company. The history of this area of technology would have been somewhat different.

Harry D. Young

1954. I went to work at the Maryland Electronic Manufacturing Coorporation (MEMCO) in College Park, Maryland, in 1954. I was assigned to Bill Perecinic and worked on various aspects of radar beacons.

1957. MEMCO's strong microwave and antenna group ordered a Luneberg lens for some experimental work. This type of lens is a dielectric equivalent of an optical lens made in the form of a sphere. The dielectric constant varies from minimum at the surface to maximum at the center. When the lens was delivered, the receiving area wanted to make sure it was in the packing box. When they opened it and found a spherical ball of foam material, they thought the lens was packed inside so they cut it apart. Luckily, they were able to use the lens by taping the halves back together.

The microswave/antenna group had a contract with Lockheed to do a radar signature analysis on a new fighter aircraft. When the Lockheed people came to MEMCO from Texas to discuss the contract, they were asked for fuselage drawings to assemble a mock up. Lockheed dissented, indicating the drawings were classified and that it would be too difficult. MEMCO said they could not make any measurements and would not be able to perform on the contract without the drawings. Lockheed finally agreed to supply them.

Several weeks went by without any contact. We finally called and asked about the drawings. The contact said, "Wes sent them out the day after he returned to Texas. Wait a minute while I check my desk. Yes, I have a classified document receipt here signed by your department."

The engineers immediately went out to receiving, searched for the classified drawings and found them. They were loft drawings in the form of four feet by six feet and four feet by eight feet aluminum sheets stamped CONFIDENTIAL.

MEMCO was having a cash flow problem. A small company on the same street in College Park, the Ahrens Instrument Company, had been bought by Litton Industries several years earlier. Our president came in on Monday morning and anounced, "We're going to hold a special stockholders meeting tomorrow night to sell the company to Litton." Since he owned a conrolling interest, fifty-one percent, in MEMCO, the stockholders approved the sale and the takeover became effective January 1, 1958.

W. A. Dickinson

As color televison expanded, monochrome televison declined. We had renewal business and sold some tubes for various displays, including coin-operated electronic games. We were saved, however, by the computer market which developed in the 1970s. We made a variety of tubes, primarily twelve-inch and fifteen-inch sizes for computer displays. These tubes were similar to monochrome picture tubes, but used longer persistence flourescent screens and traded brightness for improved resolution. The general goal was to display legibly on the tube face, an equivalent of an 8.5 by 11-inch page of typed material.

We sold tubes to the computer industry. The competition among computer makers—or those who built the display units for them—was so intense that all were very demanding on quality, especially glass, screen defects and focus. Each customer also wanted us to supply him with something unique which would give him the advantage over his competitors.

While we had many customers over the years (I worked with engineers at a number of them), my all-time favorite customer was IBM. This company did long-range planning, knew what they wanted and were willing to pay for it. They also had knowledgeable people who worked closely with us on new developments. It was no accident that the best computer display tubes we ever made were the fifteen inch, seventy degree rectangular types developed for IBM. They bought many thousands of them.

Jerome Kurshan

Professor Lloyd P. Smith was a physics professor at Cornell. During the war, he consulted for RCA Laboratories, then newly consolidated in Princeton, New Jersey, and spent considerable time there. This probably eased my own transition from student to researcher at RCA Laboratories, which I joined after graduation in June, 1943. I remember my reaction that it was much like the work I had been doing at Cornell, but at RCA I was being paid for it (twenty-three hundred dollars a year).

At RCA Labs, all technical professionals were called "engineers" in those days. That was before Bell Labs invented a more elegant term, "Members of the Technical Staff," which we and many other research laboratories adopted. Both terms, used in a unified way, have been valuable in promoting an atmosphere of relaxed cooperation. It was a far cry from my experience in graduate school where physicists had no use for engineers and vice versa, and neither had much communication with each other.

Indeed, one of the pleasures of working at RCA Labs then was the supportive interpersonal relationships. We were aware of the competitiveness, the backbiting, and the selfishness that went on elsewhere in industry and even in our own company, but we seemed to have a minimum of it. Perhaps there was some reserve between the two groups that had spawned the Princeton Labs in 1942: The Camden group under Vladimir K. Zworykin and the Harrison group under Browder J. Thompson. At the working level, however, this was bridged by the mutual respect each individual had for the technical ability of the other.

Leonard A. Karr

As a Senior Electrical Engineer with Burns and Roe (B&R), an international engineering and construction firm, headquartered in New York, specializing in power plant design and construction, I was assigned to go to Calcutta, India in March of 1962. There I was to meet with representatives of the Damodar Valley Corporation (DVC, the regional electrical utility) in company with B&R's marketing representatives, both international and local. The purpose of the meeeting was to discuss DVC's request for a proposal to review world wide tenders for furnishing and installing a central load dispatch and communications center.

The center was to provide telemetered and status information from two major coal-fired generating stations, a hydro-electric facility, and the major transmission substation. The center was, also, to include a dynamic system diagram board, economic generation control equipment, as well as teletype and voice communication facilities. Incoming information and outgoing control signals were to be carried over the exisiting power transmission system, using available, unused carrier frequencies.

Travel to India was via PanAm from Kennedy (in those days first class was authorized for long flights). A full day and two nights made this a long journey. Needless to say, I was ready for the sack when I checked in at the Grand Hotel in Calcutta.

B&R's local marketing representative showed up. He was a retired general of Calcutta. The General was used to giving orders, and the staff hopped to it whenever he demanded something, including preparation of special meal items, not necessarily on the menu. The meals with the General ordering were excellent.

The next morning, the General, Jack Kelly (the U.S. marketing representative), and I met to discuss strategy for the forthcoming meeting with the DVC people. This was scheduled for the early afternoon. The meeting with DVC went well, and, as a consequence, I was to visit DVC's major generation and transmission facilities on the following two days, in company with DVC's System Engineer.

The General hired a vehicle and driver to take us on the tour. The General also advised that we take along some lunch, including fruits that could be peeled. In fact, he ordered this for us from the hotel.

Our destination the first day was the Maithon hydroelectric facility, located some two hundred miles north of Calcutta. We travelled on the Grand Trunk Road after crossing Hoohly River into Howrah. The Grand Trunk Road had a good surface but was narrow with little room to spare when two vehicles passed each other. The practice of playing chicken was common (i.e. who would move over first). On one occasion, we came upon two buses that were locked together, neither having moved over quite enough.

At one point, we came upon a highway bridge that was under major reconstruction with an incomplete span. Undaunted, the driver, without hesitation, drove off the road, down the embankment, across the river bed (a shallow stream at this time of year) and up the opposite bank. The car promptly got stuck.

An army of would be helpers swooped down to offer assistance. Some conversation with the driver ensued. I presume he had negotiated an acceptable fee for help, since enough hand power was enthusiastically exerted to push us up and out. We arrived at Maithon rather late, but did visit the hydroelectric plant before winding up for the day.

Economic dispatch of hydro-generators, as related to other generators on the system, presents a particular problem, as the need for water management might take precedence. Water conservation during dry periods, flood control, irrigation requirements, maintenance of minimum flows—any or all may influence system economic dispatch requirements.

For security reasons and to enable quick mobilization in emergencies, living quarters were provided for the permanent staff. Facilities for transients and visitors were also provided. I was provided a rather large room, open to the outside, with a ceiling fan and (something I took particular notice of) a mosquito netting to enclose the bed. Duty at Maithon is considered a plus for its quiet, resort-like setting. A welcome respite from the humanity, hustle and bustle of Calcutta.

We planned an early start the next morning, anticipating a stop at the Chandrapur coal-fired generating station. We arrived at the station before noon. As the generator had been installed at different times, similar but customized devices would be required for each machine. The return trip to Calcutta included a stop at a typical junction transmission sub-station. Subsequent arrival at the Grand Hotel was well after dark.

The next day I met with B&R's international represent aive. He had prime responsibility for preparation of the proposal and asked me to prepare my input. His main interest, I learned, was to get the job and not to worry too much whether the job could be done at the price and terms quoted.

My viewpoint was somewhat different. I too wanted to get the job. But as I probably would be assigned to this task, I wanted to be reasonably sure it could be done as proposed.

In this case, however, there was an overriding marketing strategy involved. B&R was primarily interested in getting the larger projects of generation expansion at the two major coal-fired power plants. Hopefully, the system control center project, if done satisfactorily at minimum cost to the client, would smooth the way for the larger projects. This strategy was, of course, followed and the proposal submitted was pared to the bone. There was nothing left to do now but to return to good old U.S.A. and await the results.

As it turned out, B&R did get the central headquarters project and the two generation expansion projects. As I had anticipated, I was assigned the job of performing the central headquarters work.

Once DVC authorized the project, some eight months after the proposal was submitted, work had to begin immediately, per DVC's insistence. Hence, I prepared for departure to Calcutta in December, 1962. I was not too keen on being away from home over the Christmas holidays, but resigned myself to the inevitable.

The project, that had to start in such great haste and ruin my holiday, took many years to complete.

Hans K. Jenny

There was so much to be developed! RCA had a small engineering group at the Lancaster plant with young, dedicated individuals. No eight to five hours, we worked hard and long with much enthusiasm. New cathodes to make the tubes that would allow high speed frequency changing and modulation; new technology to provide higher quality. No routine work, all the paths were new experiences. With the end of WWII, the effort which had been devoted one hundred percent to military projects began to move towards commercial communications.

A long life, reliable C-band pulse magetron was developed for the first weather radar system for airline use. This was done together with the RCA systems division and United Airlines. This system rapidly found acceptance among the world's airlines, who today would not fly without it.

A monster 750 Kw cw magnetron went to the proof-of-design stage, but did not find a practical application. As these, and many other programs proceeded through the years, my responsibilities gradually grew through a number of positions culminating in the position of engineering manager responsible for all of RCA's microwave engineering functions.

The move from electron tube to solid state device required the on-the-job learning of new technologies while still working on the old technologies. But then an electron tube engineer always had to cover many technologies, such as electronic engineering, chemistry, materials, mechanical engineering, etc.

• Among some of the many products developed were:

• A rather complex microwave power

source for General Dynamics ARM missiles.

• A microwave solid state subsystem for a missile developed by Maxon.

• A microwave power source subsystem for IBM's WILD WEASEL program.

• Three microwave power source subsystems for the LUNAR EXCURSION MODULE (LEM) altimeter, landing speed indicator and rendezvous-radar with the mothership.

• Microwave power sources for speed radards used by police.

• Microwave power sources for the Army's handheld personnel detection radar systems.

A word about attitudes: the major profit on the electron tube business came from the sale of replacement tubes. The marketing department and mangement were orientated accordingly. Even years after solid state devices had been incorporated in newer products and systems, these people were still seeking the profitable replacement market in solid state devices which of course never materialized!

It is very hard to re-orient a once very successful and profitable organization to profound changes in technology. As a result, none of the top electron tube manufacturers became solid state device manufacturers. Needless to add that the "we've been successful doing it our way in the past" people were a real hindrance to the new generation of engineers pioneering new technologies.

Al Gross

I have the world's first hand-held radio. I built it in 1938 and it still works. The metal miniature tape recorder-sized box holds the circuitry built around vaccuum tubes. This little unit is the grand-daddy of micro-miniaturization—and I built it long before the word "electronics" was coined.

On the eve of World War II, my radio was reviewed in a technical magazine where it caught the eye of officials in Washington, DC. Subsequently, I was invited to Washington to demonstrate my design to the Office of Strategic Services (OSS)—the predecessor to the Central Intelligence Agency.

The people serving on the OSS panel must have liked what they saw. Soon I was given a commission and charged with assembling a group of people to secretly design and build hand-held radios which would operate on high frequencies. I set up my operation in Youngstown, Ohio. There I designed a two-way system that allowed OSS agents working under the net of the Third Reich to communicate directly with Allied officers flying at thirty thousand feet in modified Mosquitoes.

Called the "Joan-Eleanor" system, the units beamed a vertical signal at a high frequency—a technique which made enemy detection of the signal highly unlikely, if not impossible. Two hundred agents carrying Joan-Eleanor units were dropped into Germany. Although thirty-six of them were reported killed or captured, the rest were successful in their missions, using Joan-Eleanor systems to help bring an end to the war.

July, 1945, in a memo from the OSS to the U.S. Joint Chiefs of Staff, my system received high marks for war service, "In actual operation, it (my system) proved a valuable new tool for penetration ... into enemy territory."

Federal Communications E.K. Jett gave the idea a boost in July, 1945, when he was interviewed for an article in the *Saturday Evening Post*, "The remarkable progress achieved during the war has opened the door to a large variety of new applications of radio. One of these is the Citizens' Radio Communications Service, recently created by the FCC, under which any American citizen, firm, group or community unit may privately transmit and receive short-range messages over certain radio wave lengths."

On Sept. 10, 1945, the FCC issued me an experimental Radio Station Construction Permit and assigned the call sign W8XAG. Under the terms of the permit, I was to build an experimental radio to operate on "frequencies ... assigned by the Commission's Chief Engineer."

Relying on my Joan-Eleanor system, I submitted a prototype of a CB radio to the FCC in May, 1946. Next, I organized the Citizens Radio Corporation to develop and build the transceivers. On March 22, 1948, the radio design received the blessing of the FCC.

Although setting up the Citizen's Radio Corporation involved considerable "professional and financial risk," I didn't spend much time out on the limb. I still keep copies of these two purchase orders. One from Montgomery Ward, dated 1949, ordering $1,800,000 worth of radios. The second is from the U.S. Coast Guard authorizing expenditure of $500,000 with my company.

Remember, these were 1949 dollars and I was a thirty year old engineer. I was thrilled. A 1948 back-cover ad on *Radio News Magazine* proudly announced that the "Citizens Radio Transceiver Uses Sylvania Sub-Miniature Tubes!" This thing never would have been possible without the support and help I received from Sylvania.

In 1949, I was approached by a hospital consultant to build a "silent radio paging system for use by doctors and nurses in hospitals." Six weeks later, we had built the world's first pocket pager, or "beeper." It weighed twelve ounces and was called "Royalcall." The Royalcall also contained Sylvania vacuum tubes.

I took the Royalcall to a hospital trade association meeting in 1951, and couldn't get anybody interested in it. I was told it was impractical, that it would inconvenience the doctor and the patient. We were just ahead of our time. That's all.

Lastly, because of all the publicity I had gotten, Chester Gould, Dick Tracy's creator, came to visit. On my kitchen table was this wristwatch radio. He looked it over, and the next thing I knew, it showed up in the comics on Dick Tracy's arm.

Alfred W. Barber

In 1930, I really found myself in my element. Mr. John V. L. Hogan, the inventor of the single control radio and a former assistant to Lee De Forest had a company called Radio Inventions.

When I went to Radio Inventions, the only test equipment they had was a multi-range ac-dc voltmeter. One of my first projects was to build a TV receiver having a band-pass of 50 kHz. I designed and built the receiver. However, without some kind of signal generator, I couldn't align it. When I told Mr. Hogan, he went out into the shop and brought back a tin breadbox. "You can build a signal generator in this," he said. It didn't take long and when aligned, the receiver worked very well.

After this initial experience, I spent much of my time building test equipment. During my involvement with test equipment, I developed the modern Vacuum Tube Volt Meter (VTVM), which for many years was the standard in the field. Mr. Hogan wasn't interested so I got my own patent (No. 2,039,267) which I assigned to a company who sat on it for several years. I, however, got the patent back and manufactured the VTVM during the war as one of three qualified suppliers of VTVMs by the War Production Board. I even sold a license to the U.S. Government.

Several devices were particularly associated with the high fidelity development program. (High fidelity can be defined as that fidelity which is a reproduction of live music that is so accurate that any practicable improvement is insignificant.)

Working with Mr. Hogan, we drew up plans and specifications to submit to the FCC, with a request for a grant of a high fidelity broadcasting license. This would be the first of its kind and it would require something special, i.e. at least a 20 kHz channel. After some hearings and discussions, they granted us the 20 kHz channel at 1550 kHz. I went to work to provide genuine high fidelity broadcasting and reception. This involved analysis and solution of many problems not very important to those not dedicated to high fidelity.

An article in *Electronics* for Nov. 1935, pp. 26-29, describes some of the things I did in connection with the W2XR high fidelity project. Some components were available when I put together the W2XR's overall system—from studio microphone to listening room sound pressure—while others had to be created or improved.

Starting with the source material, namely live studio or phonograph records, I spent considerable time analyzing and improving until we had high fidelity input. I regarded and still regard live studio programs the best source. Although, there has been substantial progress with "canned" music in the past few years.

In the early 30s, World Broadcasting had a good reputation for making records especially aimed at broadcasting programs. However, before we could play them to their best advantage, we had to improve available turntables. Typically turntables were heavy pedestal machines driven by a motor through a reducing gear. They rumbled badly when one extended the low frequency response down to my high fidelity criteria of 20 Hz. I took off the very heavy platter and interposed a soft sponge rubber pad between the drive shaft and the platter itself. This worked quite well.

Next I worked on the frequency response and distortion in the records. In playing them over, I felt that some sounded much cleaner than others. I made two high pass filters, one cutting off at 5000 Hz and the other at 3500 Hz. Now when I listened to only the high frequency components, the ones with greater distortion were very evident.

I graded the records A, B and C. The A records showed minimum distortion in the region above 5000 Hz; the B records showed distortion above 5000 Hz; and the C records showed distortion above 3500 Hz. Then I made low-pass filters; a 3500 Hz cut-off filter for playing C records; a 5000 Hz filter for playing B records and no filter for playing A records. The filters were inserted with switching capabilities in the audio line of the W2XR transmitter. The operators were instructed to switch in the 3500 Hz filter for playing C records; the 5000 Hz filter for B records and no filter for A records.

When Mr. Hogan found out about this treatment of the prestigious World Broadcasting records he was very unhappy. He made me confront the World's management with my idea. They agreed I had a good idea since their records did have some problems.

It is interesting to note that back then records were generally made by engraving a wax master, gold sputtering and plating to form a die. The copies were pressed onto hot plastic of some kind. It was possible to make some very good records this way; although, there were some practical problems. Later on, and for many years, most mastering was done on magnetic tape where very low distortion is practically impossible. In other words, 1930 records could have better sound quality than later records.

I never accepted records as the best source of high fidelity. I told Mr. Hogan that if we were to claim high fidelity broadcasting, we must have some live programs. Reluctantly he agreed and I proceeded to design and build a studio at our Long Island facility.

I have already mentioned my equipment for measuring reverberation time. This W2XR studio was on the "dead" side but it was small and we had only a Baldwin piano for the sound source. We used Brush crystal microphones, the best at the time, and about the only High Fidelity microphone available (*Proceedings of The Institute of Radio Engineers*, May, 1934).

My experience with resistance coupled amplifiers at Browning-Drake and then with the 50 Hz response television signal amplifiers meant I had no trouble designing high fidelity preamps and main amplifiers. These were constructed by my able assistant, Russell Valentine. Also the modulation of the transmitter was straight-forward. The result was a very linear response and modulation from 20 Hz to 20 kHz and from zero to one hundred percent modulation.

I devised a modulation monitor using a cathode ray tube. It not only responded instantaneously so that overmodulation was readily visible, but it also monitors the signal for any gross malfunction. No meter could do either function well. It was and still is the best possible monitoring device for audio or modulated wave signals.

We never became involved with programs over the telephone lines. At one of our FCC hearings, one of their men objected to wired programs because his wife could hear something disturbing. I always felt the scientific basis for the objection should have been more than his wife's opinion.

While only indirectly concerned with high fidelity, we experimented with antennas and made signal strength measurements. In order to try a vertical antenna, I used a weather balloon to hold a vertical antenna wire. One warm day as we finished our experiments, the noon signal whistle was blown next-door. The balloon responded by bursting and letting down the antenna.

To complete the picture, I built high fidelity receivers which were compatible with the W2XR wide-band output. One of the perennial limitations of radio reception is the loud speaker. Again the Brush crystal tweeter filled the gap and we had flat response to 20 kHz. (See the article on such a speaker in *Proceedings* of the Institute of Radio Engineers Vol. 21, No. 10, Oct. 1933, p. 1408.)

Thus, I feel that in the year 1934 I completed the first integrated high fidelity system ever put together in the world. Interestingly, approximately fifty years later, an engineer writing in a high technology magazine says, inspite of fancy claims and developments, the 20 Hz to 20 kHz is really all that is necessary for the faithful rendering of musical reproduction. I repeat I invented high fidelity at W2XR in 1934 and that distinction has never been eclipsed and never will be.

Herbert Matare

During the years 1943/44, I had already proposed to use semiconductor crystals in amplifying devices which I called "three electrode crystals." Our first tests with several electrodes in varied configurations (we deposited copper points electrolytically on silicon and germanium thin films) did not work. The material was too impure and of high conductivity. For junctions we relied on Schottky contacts which did not result in sufficient barrier height and extension. The devastations after WWII interrupted all this work.

After WWII, I took up teaching at the Technical University of Aachen. However, I soon was offered a position as a research and development engineer for Westinghouse F & S in Paris. In this new laboratory, I was joined by my colleague, Dr. H. Welker, who came from Munich.

During the interviews in Paris (1946/47), I was asked about my work at Telefunken and recommended to start production of germanium diodes similar to the then popular Sylvania diodes. I was also to look for a solution to the crystal amplifier.

A new laboratory was installed and H. Welker grew germanium crystals while I started a production line for Ge-rectifiers for Westinghouse. During these years, we constantly pursued the idea of an amplifier. While Dr. Welker thought about field effect devices of the kind patented by H. Heil (1942), I tried to make a three electrode device.

Initially H. Welker grew small diameter, pencil-like Ge crystals in a Bridgman type furnace. Purity was too low to show any minority carrier lifetime. But as we grew larger size crystals, I got the expected effect with a two whisker arrangement and measured amplification for the first time. By the beginning of 1948, we were able to demonstrate the first crystal amplifiers to the French minister of the PTT, who supported our contract.

We applied for a patent and in August of 1948 a French patent was granted. The American patent was issued much later (in 1954) because of the simultaneous depositions by the group at Bell Labs. We had to consider and refer to the prior and excellent work at Bell Labs.

Right from the start I pursued practical applications. The first transistors made at Westinghouse in Paris were applied in repeaters by the French PTT. In fact, when Dr. W. Shockley visited our laboratory in 1951 he was shown working repeater models with transistors. In a summary report (1949) on our transistor work, I showed characteristics of our first transistors. Transistor circuitry was discussed in subsequent publications in German and French Journals.

At this time I was already in contact with many American colleagues and was well aware of the excellent work of the IRE in New York. I applied for membership and was accepted as senior member IRE in 1949. (In 1973, I was promoted to Fellow.)

While my first Ph.D. work in Berlin had dealt with semiconductor noise and electronics, I felt that I needed more solid-state physics training to fully comprehend semiconductors. This lead to my second Ph.D. work at the "Ecole Normale Superieure" in Paris under Rocard, Grivet et Guinnier.

In 1952/53, I went to Germany to start the INTERMETALL, Corporation, the first European producer of germanium transistors. We also started Czochralski-growth of silicon monocrystals and made the first III-V-compounds which we applied in detectors. GaSb-detectors were tested in circuits and GaAs-crystals were made. While I was building up the INTERMETALL, Dr. Welker went to Siemens where he started the well known effort to develop the III-V-materials. But problems with the financial supporters (New England Industries in New York) prompted INTERMETALL to drop all advanced work and only produce Ge-diodes and point contact transistors.

Therefore I accepted the offer to come to the U.S. Army Electronics Command in Fort Monmouth, New Jersey. Through contract work, I came in touch with many industrial outfits in the U.S. and was involved in a variety of projects at different companies, always mainly in the area of semiconductor research. As transistor electronics grew into a mature industrial activity, I worked on semiconductor defects. At the Sylvania Labs in Bayside, New York, I continued work on grain boundaries. My group constructed the first low-temperature grain-boundary field effect transistors.

Enrico Levi

In 1959 upon the death of my revered teacher, Michael Liwschitz-Garik, I had to step into his shoes, which I found uncomfortably large. Disregarding his warning about book writing that "crime does not pay," I started to write the first draft of, *Electromechanical Power Conversion.* This book was later was published in various editions, and was even translated into Russian. Perhaps because I was an electrophysicist, I tried to unify the two fields by dealing both with conventional electrical machinery and naturally occurring phenomena. At the time, the Space Age was just being born, and I thought that my engineering students should be encouraged to broaden their horizons beyond the bounds of the Earth and into astrophysics.

Much more down to earth is my second book: *Polyphase Motors. a Direct Approach to their Design*, which has also just been published in Chinese. I was motivated by the thought that machinery design was becoming a lost art, and that I owed it to my teachers to pass on to a new generation what they gave to me.

Teaching, however, has not been my only occupation. Since 1956, I was an active participant in the "Star Wars" effort and, if I contributed anything to its accomplishments, I am proud of it. Many of the things I worked on thirty years ago have been reinvented recently—and even patented by others—but this is unavoidable in a program of this scope. It is unfortunate that the publicity of the last few years has engendered a false impression about this effort. It has given us the opportunity to learn a lot of physics. Among its spin-offs are many technologies that characterize the world today and, in particular, those in which the U.S. still holds a lead.

Harold W. Lord

1946-1966. Soon after the end of WWII, the Electronics Laboratory activities were moved to the General Electric Co. Electronics Park in Syracuse, NY. I was asked to move there, but I also was offered a job in the Research Laboratory. I elected to accept the Research Lab position.

During 1946, a high interest developed in magnetic amplifiers. When the Allies investigated a German "pocket" battleship, it was found that thyratron tubes had been replaced by fast-response magnetic amplifiers for controlling the laying of guns. They were used for controlling both the motors driving the gun turrets (azimuth) and those driving the elevation of the guns.

Our Navy contracted with several U.S. industrial firms to develop magnetic amplifier systems. We were also to develop sources of the highly grain-oriented fifty percent nickel/fifty percent iron alloy which the Germans had developed for use in their magnetic amplifiers.

The development of the core material was a joint effort coordinated by the Naval Research Laboratory. Some sample ingots of the German material, which had been alloyed under vacuum, were furnished to several magnetic materials producers. A German scientist, who had joined the U.S. Signal Corps at Fort Monmouth soon after the war ended and was knowledgeable about the coldrolling and annealing procedures, was made available to assist the U.S. industry in this project. I participated in several joint meetings during this materials investigation and helped to evaluate the materials in magnetic amplifier circuitry.

In connection with my laboratory work, through the use of oscilloscope displays of current, voltage and dynamic hysteresis loops of cores taken under actual circuit operation, I learned the true relationships between the magnetic characteristics of the core material and the control currents required to control the output currents. The results of this work were published in an AIEE paper in 1953.

Rowland Medler

And the station continued to grow. We soon went into TV and the pains of that emerging technology. Man, what a hassle! But through the years of falling towers, managerial threats of bankruptcy, and moves to higher mountain top locations, the station thrived and was recently sold for probably a hundred times the original investment plus the regular profits.

On the resignation of my mentor, the Chief Engineer of WJHL/AM/FM/TV, O.K. Garland, I inherited his job. I learned much about his reasons for resignation and, after twenty years with that corporation, decided to follow his footsteps in 1958. An opening came at the

University of Florida for the beginning of a new so called "educational" TV station. I applied and went there at a considerable cut in pay as their first transmitter engineer. What I thought would be some slowdown wasn't. In a few years, while Communications Engineer of this outfit, I had my first heart attack. Thank God it wasn't permanently fatal. It became the most relevant lesson in my life, i.e. what is important and what isn't!

However, I very soon drifted back into the habits of doing what isn't important instead of, as Engineering Manager, delegating the authority and labor to those I considered least competent than myself. I goofed again and under the same circumstances would probably do it all over. I requested a demotion to Communications Engineer.

But our guidelines request a note of our greatest accomplishment. Well, our local consultant, Bill Kessler, gave me the news that there was a e5 KW TV transmitter available for the asking at WBBM in Chicago. (They had moved from the American National Bank Building over to the swank new site in the John Hancock Building.) I bit. And I worked.

I was given a week's leave to go on this small errand with the provision I make up forty hours of shift time in repayment. Actually that was a bargain, since I was regularly working sixty hours of shift time per week in addition to the office and administrative duties of the "managership." We took down thirteen tons of that old water cooled rig from the thirty-ninth floor in passenger type elevators and had it hauled to Gainesville, Florida.

The transmitter had been through two Chicago winters in an unheated area and the water cooling system had several freeze cracks. It was on channel 2 and we were channel 5. General Electric said it wasn't possible, not to mention practical, to note the channel in the field. So, complete with trusty hacksaw and torch I did it.

My budget for the project was zero. It came out of pocket or from whatever I could scrounge in material and labor from campus cohorts. On completion of the installation, I reported to the General Manager that we were ready for Proof of Performance inspection and was refused the use of a consultant engineer. So I did it myself and did the needed FCC paper work for licensing. So help me, I didn't cheat. The rig passed with flying colors and in living color, too. Thus, WUFT was for the first time a full power, 100 KW TV station.

My only comment from the General Manager was, "Rollo, a university has no business with a full power TV station." While I consider this the climax of my accomplishments, under the same circumstances I probably would N0T do it again. Or, at the very least, I would insist on repayment for my Chicago hotel bill. I might also insist that the university wheels override our manager's refusal to write WBBM's management a letter of acceptance and thanks so they could obtain a tax deduction.

Chapter seven

Dilemmas

Elias Weinberger

I was always very careful not to do anything unethical. If I thought something was not ethical, I would discuss it with the supervisor before taking any action. I can remember only one case in my career while a young engineer. I was the project engineer for the procurement of several hundred special communication devices from a small firm in the New York area.

Upon visiting the firm to determine progress on the contract, I discovered that the owner was using nonapproved parts in the unit. I informed him that this was not acceptable. After a long and loud discussion, he offered me what appeared to be a bribe. (Although, I never did open the envelope.) I left the factory, returned to Washington, informed my supervisor of the incident and asked to be taken off the project. One of my colleagues was assigned the project and I briefed him on what I had found at the plant. That particular contract was never completed! That is as close as I ever came to an unethical situation.

Charles R. Smith

My greatest dilemma was when the company (Cutler-Hammer at the time) divisionalized and my development department remained under a corporate vice president. He and the Divisional General Manager were one hundred and eighty degrees apart on many things pertaining to new product development procedures. I received all my project money for my department from the Divisional Manager. It was difficult at any one segment of time to have both of them satisfied with me and my department.

J. Rennie Whitehead

The post-war head of TRE, W. B. Lewis, had emigrated to Canada to create the Atomic Energy of Canada Research Laboratories. In 1951, he persuaded me to follow him to join the Eaton Electronics Lab at McGill University. This was an unhappy decision in one sense, for the Lab was small and its direction narrow-minded; the university faculty meetings were always full of petty arguments. It was all very parochial and it was the first time I found myself unable to run things the way I wanted them.

Shortly after the Research Laboratories were completed, the parent corporation went through a bad financial period. One day I heard that the Executive Vice President was coming up from New York to announce a decision to close the Laboratories as a cost-saving measure. At that time, we had no outside contracts, nor had we sought any. The crucial board meeting was to be the following day. We had about eighteen hours to save the Labs. I thought of all the promises I had made to the excellent team I had assembled. I felt I could not let them down.

That evening I fabricated a totally fictitious report showing that the entire operation would be supported by outside contracts starting the following month. The cost to the corporation would be nothing. My secretary typed and bound the requisite number of copies of the report that evening, obtained the key to the boardroom from her sister, who was secretary to one of the executives, and placed a copy at the bottom of the pile of papers at each place in the boardroom.

I called a U.S. friend, who alerted Elmer Engstrom (who was then Vice President, Research and Development, but later became

President of the Radio Corporation). The latter flew up to Canada and attended the meeting. He was the only person present who knew the report was there. So, it was even a surprise to the President of the Canadian Company when Engstrom called attention to it just as the hatchet man began to read the requiem on the Canadian Laboratories. The meeting broke up in confusion.

I was left only with the problem of financing the Laboratories from outside sources, which I did for the next seven years (during which time there was absolutely no further interference from management on either side of the border). When I left RCA Victor, under very friendly circumstances, the Executive Vice-President who had suffered the setback ten years before, came up to Canada and, at a management meeting, good-humoredly told this story against himself. In all the ten years I was there, he had never let on that he knew. In this case, I think the end justified the means.

Robert McLane

The most major ethical dilemma faced was my naiveté regarding office politics. I firmly believed that if I worked hard and did well, my efforts would be rewarded without conscious efforts on my part to make sure the "right people" knew about them. In one specific technical proposal I was assigned to lead, I knew, from my IEEE national committee associations, that we didn't have a viable competitive position. I verbally advised management that we "no-bid." However, I was directed to go ahead, and ask for any and all technical expertise within the company to support the effort. We made a Herculean effort to do a good job on the proposal, but our total lack of experience in the requested technology cost us the bid.

I (months) later learned that a memo had been written to my "secret personnel file" by one of the "experts" I had recruited. This person stated that I had done a miserable job in leading the effort. I then learned the importance of documenting recommendations of "no-bids!" My baptism to the games of office politics taught me that "clean living" and prayer weren't enough in the workday world of "dog-eat-dog." Upon this revelation, I learned that my career was affected more deeply by office politics than I had realized.

Edmond S. Klotz

There was an unusual chemistry that existed among the engineers at Gilfillan unlike anything I had seen before or have seen since. We genuinely enjoyed our work and working with each other.

However, we had an overzealous security director who constantly waged war with the engineering staff. He cited several staff members for security infractions because they were observed gambling on the premises (flipping coins to see who would pay for the coffee). He also issued an order that no one was to leave the premises during working hours without presenting the guard at the entrance gate with written permission from his department head. I advised the management that while it was its prerogative to require the written permission as a condition of continued employment, it could not keep me from leaving any time I wanted. I submitted my resignation but the order was rescinded before it became effective.

Jerome Kurshan

I was faced with few ethical dilemmas during my career at RCA. Those that occured involved someone higher up overriding my technical judgement. Perhaps, in all these situations, I failed to factor in the political implications adequately.

One situation arose in 1963. I was responsible for our computer services and we were preparing to step up to a powerful mainframe computer. A careful analysis of the factors i favored an IBM state-of-the-art computer.

However, RCA had entered the computer business in a big (and expensive) way and had developed its own big gun, the 601. When word of our intent reached higher corporate levels, several of us involved were called to an early morning meeting by the Group Vice President, whose responsibilities included the Computer Division of the company. He told us we were going to purchase the RCA 601 and that was that.

We did get good use out of it, although only three were ever sold. I suspect that more than just technical considerations also went into New Jersey Bell's decision to be one of three. We provided them with backup, but the computer was pretty reliable and this was not a big problem. It served us well until our next major investment, which was (can you guess?), an RCA 501.

Another case, I recall, occurred when we were ready to upgrade our internal telephone system in 1979. By then I had responsibility for a various number of services, including this one. AT&T had recently emerged from its monopoly position; having to sell its capabilities was a new experience for the company. Despite a rather poor presentation, it still seemed that its "Dimension" system was the best buy for us. The RCA Service Company had entered the business of providing "connect" telephone systems and was marketing Northern Telecom's "SL-1" system for this level of service.

This time it was the Vice President of the Laboratories who called us in and gently explained why we should go with the SL-1. Some of the expected benefits, like helping the Service Company with our research inputs, never did materialize. Partly, this was because the business potential evaporated when the manufacturer decided to take over the marketing of this system in the U.S.

Fred Tischer

My "space adventure." It was a cold and wintery day in January 1962. I was teaching in the electrical engineering department of Ohio State University (OSU) when I got a call from the University telling me that a representative of NASA wanted to see me.

"Space" was, at that time, a really hot topic. It quickly became a crystallization point attracting outstanding and forward-looking people from all over the country and from abroad. The universities drastically changed their curricula. The industry introduced quality controls, and the reliability of merchandise became an important requirement. The early sixties became a time, during my career, when quality and perfection were really highly appreciated. This trend permeated all aspects of life in the U.S. and it became the intangible basis of successes in the space activities with the landing on the moon as the miraculous climax.

After a brief period of negotiations with NASA and OSU, I moved temporarily to Washington, DC. My activities at the Goddard Space Flight Center were associated with plasma and re-entry communications. They were comprehensive and exciting. The availability of scientific and technical information was unique; conferences with scientists and visits at various laboratories and industrial plants were arranged instantaneously. The work was extremely stimulating and the environment very supportive.

At the end of September, with my return to OSU seemingly imminent, I was asked to visit the Marshall Space Flight Center to meet Dr. von Braun and one of his former associates. The gentleman offered me the position of Assistant Director of the University of Alabama Research Institute in Huntsville. (The Alabama State Legislature, on a request by von Braun, had appropriated four million dollars for its creation. On the airport, at that time, you could read "Huntsville, the Space Capital of the Universe." Von Braun now had the idea to create "The Space University of the USA.")

Von Braun said I would be in charge of radio, wave propagation, plasma, and communications. It was a fascinating project, and I indicated my definite interest in the position, with the proviso that it be an associate directorship instead. During negotiations things were fouled up and when I received my appointment letter from the University of Alabama, it was for Assistant Director and Professor of Electrical Engineering. Since I was already in Huntsville at the time, I accepted the appointment with reservations.

The two years I spent at NASA and in Huntsville were very exciting, particularly the time at NASA when I made important contributions. I would not have liked to miss them. The high level planning meetings on the Apollo project were memorable events.

An annoying factor during the Huntsville days was the politics involved in the operations of the Research Institute. The Director, who had played a double-faced role at my appointment, experienced a similar fate. After a year of operations, a man who was supposed to become Business Manager of the Research Institute appeared one day carrying the title, "Vice President of the University of Alabama at Huntsville." The Director, who had been the top man, unknowingly was demoted one step down the hierarchical ladder. It also was becoming apparent that "The Space University" was out; operations were being relegated to an "extension of the University of Alabama at Huntsville" status with the curriculum tailored to junior college operations. Several of the scientists/professors subsequently left, including myself. And that ended my "space adventure."

Nick Petrou

I firmly believe ethics cannot be taught as a separate course in graduate schools. In actual practice, ethics is an every day issue and an integral part of every day living. Ethics begins at home and progresses throughout one's educational life.

Franklin Offner

I believe there should never be an ethical dilemma for an engineer or scientist. The only acceptable course is complete honesty, without reserve. When I found that the company promoting my EKG was less than completely honest in its representations, I immediately left. I found it necessary to do the same after having merged my company with Beckman. I have found that the most ethical companies are usually the most successful.

I have always worked either independently or as a team leader, though the objective of my work may have been dictated by others (my development of the EEG for Professor Gerard's laboratory at the University of Chicago, for example). For the projects which were under my control, I had the full cooperation and back-up of United Aircraft and the Air Force and, of course, I always had the facilities and personnel of my own company at my disposal. However, this all changed radically when my company merged with Beckman. Innovation became almost impossible.

When I joined Beckman my new ideas were almost all automatically rejected. For example, when I invented the tunable monochromatic laser for spectral analysis they considered the idea worthless; I later sold the patent to another company, from whom Beckman must now purchase the lasers.

Robert Newman

We continued to have difficulties with General Electric's company edicts, particularly with the accounting and cost conventions. Once I convinced the "design engineers" to design a tube line using modules and common parts to reduce manufacturing, inventory and tool costs. The first two tubes in the line had a before-tax profit of twenty percent—better than the average product we were making. However, the accountants insisted on costing the two separately, and reported that the larger tube had a forty percent margin and the smaller one a two percent loss.

Each year the smaller tube was on the "improve cost or dispose" list and was finally dropped. A year later the larger tube—having lost economies of scale—was on the dispose list. Thus, this once promising line was abandoned because we dared to use a manufacturing theory not compatible with an accounting convention.

Frank K. Faulkner

My sixties included participation in a team effort to improve WAPDA's electric distribution system in Pakistan. Although this should have been the high point of my career, complications including poor cooperation of joint venture partners, in transience of Pakistani nationals, and redirection by USAID officials resulted in my disappointment and retirement. An ethical conflict resulted when the Asian Development Bank (ADB) was assured by USAID that the joint venture team would assist in the thirty million dollar ADB distribution loan project. However, the joint venture team was restricted to work only on the USAID fifty million dollar project.

Paul G. Cushman

The Polaris Project was managed by the Special Projects Office of the Navy. MIT received a contract for the guidance, General Electric (GE) received a contract for submarine based fire control and guidance support, and Lockheed received a contract for the missile structure and propulsion. These were intially development contracts, which were extended to produce hardware. (There were other contracts—for example, submarine, nuclear power plant, and submarine navigation—but these did not interface with GE directly.)

Overtly, MIT, GE and Lockheed cooperated harmoniously. Covertly, there were rivalries. Liaison representatives were exchanged between the three groups. Usually such representatives would attend meetings and take care of their correspondence. But the GE liaison man at Lockheed, Bob Howell, insisted on taking a more active role than that, and was soon running Lockheed's large scale, analog missle trajectory simulation. His findings from these runs were important facts in the system and hardware decisions that were made by Lockheed. In this "triumvirate," it appeared unlikely that I, as Inertial Guidance Concept and Systems Engineer, would be allowed to make any significant contribution to the project. As it turned out, I was able to make a contribution, but the acceptance of my contribution required the manipulation of the "politics" of the situation. To a certain extent, my methods might have been labeled unethical, but it didn't seem so at the time.

The guidance steering equations that were proposed by MIT could not be used early in the missle flight. Their use had to be delayed until the missle had been pitched over and had acheived considerable velocity in the general direction of the target. MIT proposed that the early part of the trajectory be achieved by a missle attitude angle program. However, Lockheed soon showed with its trajectory simulation that expected wind conditions could cause mechanical overloads on the missile struture if a fixed angle program were used. So then world wide winds began to be investigated regarding their predictability so that a highly variable set of angle programs could be generated to be used at different launch locations and directions of fire.

This kind of complexity and uncertainty did not please the Navy any more than it did Lockheed. With this dilemma in hand, I spent a few days modifying the steering equations used at the terminal portion of the Hermes A-3 trajectory to the boost part of the Polaris trajectory. The concept utilized the signals from the integrating accelerometers mounted on the IMU stabilized platforms to acheive a highly damped, lateral velocity control for the missle. In such a control, any wind loads are sensed by the accelerometers and automatically compensated by the control response. The calculations that I did were considerably more than "back of the envelope," but needed simulation trails for evaluation.

As industrial supporters to MIT, ideas like these generated at GE should have been disclosed to MIT. However, on the basis of previous contacts with MIT, I knew that ideas from outside that organization were not sought after, and, if offered, would be ignored unless presented with a great deal of supporting data.

So I suggested the idea to Bob Howell for trial on the Lockheed Simulation. The results were immediately favorable. Not only were the required implementations much simpler than even the single, fixed-angle program systems, but the simulation showed that the missile could fly through the worst expected wind profiles with no trouble. As a result, Lockheed forced MIT to incorporate Z-Dot Steering (my nomenclature) into their guidance concept, and this steering was used on the early Polaris missiles.

It was a little amusing to me, that nearly thirty years after the concept of Z-Dot Steering for flying though wind profiles had been demonstrated, a shuttle launch was delayed several hours because a "winter wind profile" had been programmed into the guidance computer and the wind on that day was blowing from the opposite direction. It is often said that engineers are prone to "re-invent the wheel." Probably an even greater failing is their tendency to substitute an inferior invention for the wheel.

M. Lloyd Bond

In the late '40s, my work was generally in engineering design under military contracts—with the final job (of the '40s) as Project Engineer on a missile guidance system. Unrealistic deadlines led to enormous amounts of overtime—all unpaid. When the group which I directed complained, there was a discussion between the plant manager and myself which led to the comment, "If you make an official complaint about unpaid overtime, you'll get it, but you will not have a job any longer."

There were also major shortcuts in quality control tests resulting in improper release of missile units. It was necessary to resort to personal calls (off hours) to the government's Project Manager. He then just "happened to" drop in to witness tests of units about to be released for flight testing. Whistle blowing was averted this way.

The sixties saw a major change. I became a senior appointed government official in Communications (technical) and was responsible for the design, execution, and operation of the Federal Telecommunications System (FTS) and the Advanced Record System. The civil service pay was far less than what I had been receiving before joining the government, but the training, responsibility, and challenges were the best I had during my career. After Kennedy's death, the challenges disappeared since LBJ's government was predominantly politically (and not technically) motivated.

In the latter half of the sixties, I gravitated towards marketing administration—clearly technical and needing engineering background, but not of a design nature. I had become too "old and dated" in my learning to continue in engineering or engineering management. The money in marketing administration was very good, however.

The FTS system has to be the landmark of my career while the experiences in training while in the Navy, the worst. Yet I enjoyed engineering so much that I probably would not have done anything differently (except modify my aggressive personality). The last seven or eight years before retirement were frustrating. I wanted to use my technical expertise in communications more than being a vice president of marketing would allow. But age was a factor. Discrimination? Yes. Subtle, but there nevertheless.

Sidney Bertram

In 1957, with the training program running, I joined Ramo Wooldridge (later the

Bunker Ramo Corporation). In 1960, I became involved in a program to automate the extraction of topographic data from aerial photographs. This culminated in 1964 with the delivery of the first Universal Automatic Map Compilation System, or UNAMACE.

Each UNAMACE had two identical precision tables, each capable of holding up to nine-by-eighteen inch diapositives (glass positives). These were computer controlled, with the positioning commands going to both the table servos and the flying spot scanners used to examine the diapositives so that, for the small motions required during automatic operations, the images of the scans move instantaneously to their computer-commanded locations on the diapositives. The system therefore operates at the two micrometer table accuracy "on the run." The computer also controls the size and shape of the scanning rasters, making operations with unconventional photography possible.

Using conventional vertical aerial photography (six-inch focal length), the equipment measured ground elevations to about 1/10,000 of the altitude the photographs were exposed at. The equipment made about fifty measurements per second except in very steep areas. The original equipment had two additional tables that were used to prepare new photographs, with the detail of the originals but with the parallax distortion removed, and altitude charts; the altitude data is now stored digitally allowing the outputs to be prepared off-line by simpler equipments.

I am quite blasé with respect to newspaper stories of high priced toilet seats because I have seen more significant problems with the military procurement system. For example:

We were the only qualified bidder for a full-map-sheet size table for UNAMACE, but lost out to a low bidder; the military never got a table that worked.

Following a request from the user agency when the UNAMACE became operational, I applied for the clearance required to work with them. I never received it, although I previously had other high-level clearances. I am certain it was because a company doing mathematical work related to UNAMACE for the agency didn't want me to have it; their president had been with the agency.

We had another problem with classification: I presented a paper on the operating principles of UNAMACE in Portugal in 1964, and a second paper on its operation in Switzerland in 1968. Both papers were published in the *International Archives of Photogrammetry* and in *Photogrammetric Engineering*, so they were readily available. The customer and user agency also presented papers on the results of their operations with UNAMACE. When the equipment became operational in 1964, the customer apparently realized that it was much more effective than anticipated and classified certain aspects of the instrumentation as "confidential." As a result, I was allowed to give out copies of either paper, but not both together! (The second paper referenced the first.) I continued to be notified about continued classification until about 1980, though a friend who was working with the agency insisted that UNAMACE had been declassified earlier!

Changes in your career path: 

J. Egbert Sousé

Chapter eight

The road traveled

Charles T. Morrow

In a year's time, partly because I had considered the shock and vibration work to be only a one-year assignment, I left Hughes Aircraft for the Ramo-Wooldridge (now TRW). Although I was not to know it for a month or more, I would participate in the U.S. Air Force Intercontinental Missile (ICBM) program, which at that time was completely secret. (A national emergency response by President Eisenhower to intelligence that the Soviets were ahead of us by about five years.) As it turned out, although my initial assignments were in inertial guidance in accordance with my wishes, I had not seen the last of shock and vibration.

The Guided Missile Laboratory of the Ramo-Wooldridge Corporation, under Dr. Simon Ramo, or Si as everyone called him, was the technical arm of the Air Force for the ICBM program—responsible for systems engineering and technical direction. Si was a virtuoso at the type of management required. He typically delegated everything completely, but monitored the progress and did not hesitate to step in if there were any signs of trouble. He made effective use of the most knowledgeable consultants available. He emphasized good communication, so we were not merely a disconnected group of independent specialists trying to do systems engineering. There was a colloquium every Friday morning, attended at the beginning by all civilians and Air Force officers who were in town and in good health.

Understandably, Convair in San Diego, which had privately funded an ICBM investigation for several years, resented the prominence of Ramo-Wooldridge. Many of my colleagues encountered difficulty at Convair, but I never did.

Probably the first action taken by the Guided Missile Laboratory, later renamed the Space Technology Laboratories, was to downsize Convair's concept of the ICBM in accordance with the lighter nuclear warheads that had become available. This made the missile easier to develop and more transportable. It resulted in our having less weight lifting capability when the space age arrived, but otherwise, we might have had no operative boosters in time.

There were high risks in all subsystems of the ICBM. For example, we did not know for sure that the warhead could be protected to survive the heat of re-entry. (Presumably the Soviets had solved this somehow.) Consequently, we did not know for sure what the size of the missile had to be.

In all areas, we made guesses on the basis of the best evidence and judgments available, and took our chances. Alternate approaches were followed simultaneously for most subsystems. Before long, the participants were divided into two teams, Atlas and Titan, with an initial requirement that a subsystem that proved necessary for the other missile could be adapted in a specified minimum time. All approaches worked. The radar guidance systems proved indispensable for the space age.

There comes a time in a missile development when test firings must be made even if the chance of successful flights is low. Otherwise, important problems may not be discovered until late in the program. Accordingly, the Atlas was undergoing firings in Florida without guidance, but with enough control to keep it from tumbling. Again and again it blew up on or above the launch pad. Eventually the problems were straightened out and guidance was installed and tested along with the rest.

(The initial history of Titan firings was different—the first six firings were successful. The engineers became very worried. Reliability could only go down, and they would be asked why. It did go down and they were asked why.)

I was a member of the Electronics Department, which, under Dr. James (Jim) C. Fletcher, son of Harvey Fletcher of Bell Laboratories, had responsibility for guidance, control, tracking of ICBM's in flight, and related electronics. I had known Jim at Harvard and at the Hughes Aircraft Company. He was as enthusiastic about good communication as Si, well liked, and already respected almost as much. Since then, he has twice been Administrator of NASA—the second time over his public protests.

After I had been at the Guided Missile Research Laboratory for several months, Jim asked me to prepare an environmental specification "for the Atlas." (Division of the effort into two teams had not yet taken place.) This was suggested as a weekend project to give the associate contractors an idea of what to look forward to, but it did not turn out that way.

No sooner did my draft become known within GMRD, than I became subject to attack by William (Bill) Besserer, by reputation an excellent administrator. Undoubtedly, this started as an organizational dispute—it was common in the aircraft industry for structural (airframe) engineers to establish the shock and vibration conditions for airborne equipment.

However, the attack immediately focused on the technical aspects of what I had prepared, especially my innovations. These included a random vibration test specified in terms of a spectrum of power spectral density of acceleration, a shock test specified in terms of a minimum shock spectrum, and the suggestion of a sawtooth pulse of acceleration, with appropriate test apparatus, as a means of achieving the shock requirement. Bill's reaction to my involvement in shock and vibration specifications was understandable, but I had no authority to withdraw.

In retrospect, he played his hand of cards exactly wrong for his own interests. If he had the patience to let my draft go out to the contractors as originally planned, the resulting confusion might have been so great that he could have stepped in with an alternate document that would have been accepted. But he forced the debate to take place where the engineers knew little about environmental testing but much about random noise theory and spectra, instead of much about current practices of testing but little about spectra and random vibration. A year later, after a frustrating battle, the specification was issued with little change in the innovations. However, it was no longer just a rough draft. It had become an official specification, approved by both the Ramo-Wooldridge Corporation and the Air Force.

The only repercussions came from the MIT Instrumentation Laboratory, which was developing gyroscopes and accelerometers for inertial guidance. (However, these did not reach official channels.) I was sent to the MIT Instrumentation Laboratory to confer.

When I arrived I was ushered into Director's office, Dr. Stark Draper. I was introduced as the man "who had written the environmental specification for the Atlas."

Draper exploded. He ranted and raved—it seemed for an hour or more. Finally he got around to the aspect of specifications that bothered him most—temperature cycling. When eventually he paused long enough for me to speak, I commented, "I did not include a temperature cycling requirement in the specification—maybe I should."

The tirade came to an abrupt end, and I was able to go about my business. I may be the only person that got the last word with Stark Draper. I did not introduce a temperature cycling requirement. I thought I had better leave well enough alone.

Curiously, the specification had more impact on the industrial structure of the United States, and in particular of Texas, than I could have foreseen. The only company in a position to supply the required amplifiers, LM Electronics, was in financial difficulty. The company needed to be purchased to obtain stability. Finally, it was bought by an almost unknown Texan by the name of James Ling. He renamed it, Ling Electronics, and made it the cornerstone of the Ling Temco Vought (LTV) empire.

At the Ramo-Wooldridge Corporation, as all this was in its infancy, I was a member of the Reliability Working Group, a committee initially chaired by Jim Fletcher. Missile reliability had been a generally sensitive matter for several years. The statisticians in the group favored assignment of subsystem and component reliability "objectives" based on the somewhat unrealistic assumption of statistical independence of failure rates. They accepted this as an axiom. They demanded verification of the objectives under "specified" test conditions. They were not concerned about systematic error or cost penalties. The latter, even though no budgetary limit had yet been applied to the ICBM program, were clearly so high that management could not approve systematic verification.

I devoted my energy to finding constructive ideas that management could approve. The statisticians reacted to this with all the self-righteousness and horror of a fundamentalist revival meeting—as if I had made a speech in favor of sin. Management finally approved the assignment of reliability objectives but not their verification.

After about a year, Duane departed in poor health. I inherited his responsibilities. One was the annual Symposium on Ballistic Missiles attended by the Guided Missile Laboratory, the Air Force and the associate contractors. Duane had directed this entirely by himself, except for the assistance of his secretary, Caroline Budelier.

I directed the third Symposium much as Duane had done it, with the aid of Carolyn, who supplied me with essential information. One problem was that the Air Force regarded this merely as an in-house meeting. If at the last moment an officer wanted one of my conference rooms for a meeting, he took it. After this one experience, I determined that I would not do it the same way again.

Unexpectedly, the Soviets helped me! After the satellite Sputnik I was launched, I proposed that the Symposium be moved to UCLA, unclassified sessions on space be included, and the press be invited. The Air Force approved. For several years, this was THE national meeting on space.

Although my responsibilities were almost

entirely administrative, I kept up some technical research on a noninterference basis. This was to be my salvation years later. I presented and published about one paper a year, usually on shock and vibration.

The random vibration concept was adopted outside the ICBM and space programs, but not until after ten years of violent controversy. A more reserved debate on shock spectra and swath shocks continued for thirty years before their general adoption. The official approach to shock spectra remained less than optimum, partly because computers, even analog, were not generally available when I had transferred the concept from the Navy to the Air Force. Consequently, the specification requirements had to be written in terms of what a mechanical instrument, the reed gage, could do. Then the shock spectrum computers were designed to do what the specifications required.

I attended almost all the meetings of the Shock and Vibration Symposium, then sponsored by the Shock and Vibration Information Center (SVIC) at the Naval Research Laboratory. It was at this Symposium that I learned much of what I needed to know about shock and vibration. SVIC eventually became my client.

The outstanding success of the Guided Missile Research Laboratory, since renamed the Space Technology Laboratories or STL, prevented the rest of the Thompson-Ramo-Wooldridge corporation from competing for missile contracts. Accordingly, the Aerospace Corporation was set up to take over systems engineering and technical direction. The Symposium on Ballistic Missile and Space Technology was transferred, and so was I, along with others who were to set up the new corporation.

I now reported to Dr. John (Blackie) F. Blackburn, a widely respected tenderhearted man with often a gruff manner. He had long been associated with the new president, Dr. Ivan A. Getting, at Raytheon where they both had staff responsibilities, and earlier at the MIT Radiation Laboratory. Blackie had a photographic memory, but he seemed uncomfortable making decisions. When I needed a decision, I would state a problem, then listen patiently while he enumerated and discussed all possible options—until he came to the option I favored. Then I would enthuse. This was usually sufficient. Blackie seemed to be pleased with my practice of running my shop without bothering him unless absolutely necessary. His title was Consultant to the President.

I had somewhat more official status now at the Aerospace Corporation, but less authority than before the transfer. Previously, with the gentle guidance of Vice President Ralph Johnson, I spelled out policy on some matters. Now I formulated no policy except by subterfuge. On occasion some management person would try to set up unworkable restrictions on clearance of papers. I managed to dodge these. From my vantage point, the corporation seemed to be run more by consensus than by leadership.

Blackie seemed to be in a continual state of frustration. On occasion, Ivan would invite the two of us to a management meeting, ask our opinion publicly, then come up with one of his own that was obviously superior. Of course, he had a vantage point that was better than Blackie's, and much better than mine. Also, he may have been more brilliant than either of us.

But he tended to compete with his staff rather than use it to help staff. Although I witnessed additional signs of dissatisfaction within the management, I thought working conditions at lower levels were probably good. But eventually, after I left, the engineers established a union.

The Symposium on Ballistic Missile and Space Technology was brought to new quality and distinctiveness for two final years by a heroic effort at the planning and invitation of papers. This involved some risk. We could get stuck with poor papers which we could not reject. But we did well. Then the meeting was done in by friction within the Air Force.

There were at that time a Ballistic Systems Division, in San Bernadino, and a Space Systems Division, in El Segundo. In a peculiar organizational arrangement, the latter was designated the lead division, which caused some resentment in the former. The former took the position that attendance quotas for the Symposium should be limited, reducing the total attendance. The latter did not have strong feelings but finally agreed, thereby establishing a policy.

Then a Space Systems Division colonel, who had been a major asset to the Symposium, suddenly began to think of it as his own meeting. He would visit an associate contractor, talk to the president, and promise to accept an unlimited number of attendees, then request by telephone that I honor the promise. My office made note of such requests, but, under my orders, took no action. Otherwise, I would have been violating Air Force policy. I took the risk of anger from one Air Force colonel; but otherwise, I would have had at least one general and at least one Air Force division mad at me. Of course, I kept Blackie informed about what was happening, and he presumably informed Ivan.

Before long, my first confrontation with the colonel took place. I remarked that I did not take verbal orders. He actually wrote out an order for me. Within two hours it was on the desk of his commanding general and he was up for reprimand. He may have lost whatever chance he had of becoming general, but the reprimand did not stop him. I do not remember the details of what followed, but he showed up on the war path, two successive Mondays, and had to be knocked down again. In the course of this he high-handedly antagonized the Ballistic Systems Division, which registered a protest. The Symposium took place successfully on schedule at the Air Force Academy, and I emerged unscathed, but further meetings of the series were canceled.

At that time, the Aerospace Corporation had no provision for adjusting departmental quotas to permit a personnel transfer, except for management at least one level higher than mine. I recognized that I would have difficulty staying at the company beyond the completion of the proceedings.

Finally, I accepted a position at the LTV (Ling Temco Vought) Research Center, Western Division, in Anaheim, California, directed by Dr. John Hilliard, one of the more durable pioneers of acoustics and theater loudspeaker design. Although I had enjoyed many aspects of administration, I now enjoyed getting back into acoustics and electronics, and I enjoyed my new colleagues. Shortly after John's retirement, LTV encountered financial difficulty. So did Ling Altec and Ling Electronics nearby in Anaheim. LTV decided to close down our laboratory and to invite me among others to Dallas.

After three quarters of a year of apparent indecision, on Friday, the thirteenth of November, I started a long cold drive to Dallas—ironically, just in time for Jimmy Ling's demise. In this way, I was in on his beginning and his end.

Within two weeks after arrival, which was of sufficient importance to be announced in the LTV newspaper, I determined that the chances of the LTV Research Center lasting were no better than fifty-fifty. My reason was simple: there was essentially no chance of the owners utilizing any of the research results. Jimmy Ling had not assimilated the companies he had acquired.

Many of my colleagues complained about the quality of management in the research center. I did not complain, for I knew in the long run, quality of management would make no difference in the future of the laboratory. But they had hope; I did not. However, as one of two staff scientists, I had one of the best positions there.

I was laid off one year before I would have vested retirement. Soon afterward, the research laboratory was absorbed into the airplane factory. As LTV went bankrupt, I do not know whether vesting would have benefited me or not

I did some successful consulting in Dallas and later in California after our return. Inflation and falling stock prices, during the Carter administration, made our savings almost worthless. For two or three years, everything I tried to do professionally encountered a road block. The Navy did not have the money to publish my study for SVIC. I signed a contract with the president of Diving Unlimited International (DUI) of San Diego to reduce the diver microphone to a producible design. I would receive a share of the profits. The comptroller, who had not been consulted, pointed out that the diving equipment business was seasonal. It would be better to delay my work for several months until some other company projects were completed. By that time, DUI, on the verge of bankruptcy, sold its electronic department to Sub-Sea Systems of Escondido, California.

My work began, but the DUI electronic communication system was discontinued. Either because it was poorly designed; or because Sub-Sea Systems did not know how to build it and help the customers. Fortunately, by this time, my contract contained a deadline for production, which if not met would permit me to take the microphone project anywhere else I chose. People were assigned temporarily to the project, then reassigned—a sure way to waste company effort and my time. The technical problems were solved by the deadline, but the company made a decision not to produce the microphone. I had not taken it elsewhere, as yet, since I had other things to finish up that I considered more important. But this was the way my fortunes went for a while.

Finally, I approached retirement age in better financial shape than I had expected, and my professional fortune changed. The Navy found the money to publish my report. I revised it completely, changing the emphasis and eliminating some material I considered unimportant and distracting. The new title was, *The Environmental Specification as a Technical Management Tool.*

This report showed that technical considerations should not be the only basis for evaluating a specification. By focusing solely on testing, specifications had concentrated shock and vibration engineers into test areas where they had no authority to influence design. But detailed design requirements would have disastrous effects. The study also showed that beyond a certain point, emphasis on testing and on comprehensive design verification could have adverse effects by diverting money and effort from design. I expected some vehement criticism for the report, but I did not get it. The general reaction seemed to be that I had written some things that should have been written years before. There were no restrictions on the distribution of the report.

Leonard R. Malling

(1933-38) As active chief of Baird Television, Captain West was at this time working on high definition television, sixty lines at twenty-four frames. I visited him and told him I would study the progress of television in the U.S. and report on my return to England. By inventing a new method of film sound recording, Captain West, working for Gaumont British, had broken the stranglehold of Western Electric on the English movie making business. Gaumont British decided that there was a future in television and poured funds into the Baird Company under the direction of Captain West.

Baird had a contract with the BBC (British Broadcasting Corporation) to broadcast television, at that time sixteen lines at twelve frames, mechanically scanned and reproduced. Strangely enough there were several thousand receivers that had been sold by Baird to watch these broadcasts, mostly operated by amateurs. Sir John Reith, czar of English broadcasting, made the announcement that television was but a toy, would always be a toy and that the BBC would not waste anymore time, money and facilities for this toy. The BBC thereupon broke the Baird contract, tore out the Baird equipment, and threw it out on the street.

Captain West then moved the laboratory to the Crystal Palace—the towers there being the highest point in London—and commenced experimental high definition TV broadcasting. This was initiated with one hundred and eighty lines at twenty-four frames, using flying spot mechanical discs. The results from movie film were excellent, less so for studio broadcasts.

Returning from the U.S., which was just recovering from the depression, my task was to design practical receivers for the home. Lacking a camera, Captain West decided to acquire rights to Farnsworth TV systems. The Farnsworth camera would provide pictures to two hundred and forty lines. I became interested in the Farnsworth reproducing equipment, considered totally impractical for home use by our engineers.

We hired a phosphorous expert from central europe and I proceeded to design short all- magnetic scanning envelopes that were a first in the TV field, permitting direct TV tube face viewing. I also commenced experiments on high voltage from the scanning system. Baird was now selling TV receivers to the general public and broadcasting TV programs mostly film from Gaumont British.

I was well paid on two-year contracts and in complete charge of all TV receiver activities—the highest management level of my career. In the U.S., RCA was still bogged down by the lack of a practical TV camera. The Iconoscope at this time had too many operational problems. My responsibilities at Baird together with my educational degree permitted me to be elected as a Member of the British IEE.

Sir John Reith of the BBC now announced that the BBC had chartered rights to all radio broadcasting in England and that included TV. Having political clout, he had this confirmed by Parliament. Public TV broadcasting from the Crystal Palace was now illegal. The Marconi company, which had a vested interest in the BBC monopoly, quickly concentrated on TV using RCA technology, EMI staff and the Iconoscope.

Baird was out. A tremendous fire at the Crystal Palace then totally destroyed the Crystal Palace including the Baird laboratory and its studios. I decided to emigrate to America.

I took a job with the then fledgling Varian Associates in their small plant for making klystrons in San Carlos, California. Dorothy Varian would make lunch for the workers in the small cafeteria. I would play horseshoes with Russ Varian during lunch hours.

I built a few machines that improved the quality of klystron production. But what I really became interested in was NMR, nuclear magnetic resonance. Working with nearby Stanford researchers, such as Felix Bloch, father of NMR, I developed practical saleable NMR equipment. I published a couple of papers in *Electronics* covering this field. With the death of the two Varian brothers, Sig and Russ, I lost interest in Varian and took a job at Hoffman Radio in Los Angeles. I had been invited down by a friend I had met at Convair. I was able to make a real dent in the technology there, in the UHF frequency region, resulting in a successful military contract.

Desiring to get back into research, I went over to the JPL (Jet Propulsion Laboratories) in Pasadena and found to my surprise that I was quite well known there because of my IEEE paper on Radio Doppler. JPL was under contract to the Army. Von Braun was developing rockets at Huntsville (AL). JPL was responsible for guidance. Their guidance system worked at UHF with cumbersome cavities designed from old MIT WWII data. I redesigned them to be ruggedized and greatly reduced the size and power requirements. I published a paper on the techniques involved.

With the advent of Sputnik, JPL was called upon to build an instrumented satellite that would be launched atop a rocket developed at Huntsville. I was given the job of designing and building the electronic assembly in the rocket, comprising instrumentation, transmitter and battery assembly. Dr. Pickering allocated eighty-one days for completion. I had a team of technicians and the JPL machine shop at my disposal.

Here was my big chance to do something significant for the U.S. The satellite must work. I, of course, knew nothing about the outer space environment. I would have to learn and use my imagination. Late on the eighty-first day, I checked the weight and shipped the satellite to Florida. (I had run a continuous weight analysis to be sure that at the end the assembly would meet the precise figures.) The mission was successful. The U.S. now had its first satellite in orbit. Several similar satellites were launched. JPL became a laboratory for the newly created NASA agency.

Attention was then directed to a satellite

that would penetrate deep space and fly past the moon. I was allocated the radio transmitter with a weight limit of one pound. Only a marginal increase in rocket power was available hence the overall size and weight would be limited. I decided to push for a UHF transmitter, 960 mc using the latest technology and a team of technicians. Using transistors to 320mc, a varactor multiplier from the Bell Labs (never used in any application), and a tiny GE vacuum tube, I managed to squeeze out a quarter watt of rf power.

Horrendous problems of every type dogged my progress, but I was able to complete the transmitter within the six months allocated. It acheived its objective and was contacted for one million miles—another U.S. first. The Van Allen belt was discovered with this space probe. I was invited by the Japanese to give a paper in Tokyo, they having read an article I had published.

In 1965, I no longer wished to work on these missions. I flew to Boston, Massachusetts to get myself a job at the Lincoln Lab.

Alexander Lurkis

In 1934, Mayor Fiorello LaGuardia pulled a stunt for which I never forgave him. In the height of the depression, when professionals were selling apples to get food for their families, he arranged for the layoff of over five hundred engineers within the Board of Transportation. I was one of them. We were re-hired four months later with twenty-five to fifty percent cuts in salary. It was a budget gimmick used to get federal funds for needed work, and which gave the mayor an opportunity to claim he cut the city budget by twenty-three million dollars.

As a result of the layoff and cuts in salary, and the additional threat of public engineering and architectural design being farmed out to outside consultants, a few of us got together representing different city departments. We discussed our problems and decided that we must organize to protect our interests. We formed the first civil service technical organization and called it the Civil Service Technical Guild. I was appointed Legislative Chairman and became a lobbyist because our principal efforts required getting legislative acts to provide higher salaries, pensions and improved working conditions.

Later in the fifties, we joined with the CIO. and in 1958, we became part of the AFL-CIO. I was elected Vice President and, in 1956, became the President and a Vice President of the District Council 37. I served in these positions, after working hours, while carrying out the responsibilities of my job. The Guild achieved a great deal for the city's technicians. Today it has six thousand members. In 1987, the Guild celebrated its fiftieth anniversary.

In 1959, Mayor Robert F. Wagner appointed me to the civil service job of Chief Engineer of the Bureau of Gas and Electricity in the Department of Water Supply, Gas and Electricity under Commissioner Armand D'Angelo. For a short period in 1961, I was Acting Commissioner, while Mr. D'Angelo was on a leave of absence. As Chief Engineer, I controlled the electrical life of the city. In this capacity, I became involved in rate studies, standards for utility overhead and underground facilities, the preparation of the Electrical Code and approval of all electrical installations in public and private structures.

I was in office only eight months, when the first major Con Ed 1959 blackout hit the city, at 2:56 p.m., shutting down five square miles of upper Manhattan at the beginning of the rush hour. Service was not fully restored until 3:42 a.m., the following morning. Hundreds of thousands were trapped in elevators and subway trains; industry and commerce ceased operations; traffic was snarled as signals went blank; hospital operations stopped in the midst of surgery; water pumps failed. The city was in a frenzy.

Con Ed claimed an "act of God." Mayor Wagner directed me to investigate this happening. My conclusion was that "God" could not be blamed. The fault was in the distribution system. After the blackout, Con Ed stated that, "The mathematical chances are negligible, that a similar situation will develop again." However, Con Ed proved to be a poor prophet as evidenced by the blackouts of 1961, 1962, 1965, 1971, 1977 and 1981.

Later in 1981, I wrote, *The Power Brink-Con Edison - A Centennial Of Electricity,* which explained the reasons for each blackout. I self-published the book under ICARE PRESS. (See *Spectrum*, April 1984 and *The New York Times* 12/20/81 for book reviews.)

In 1964, I had thirty-five years of city service. I decided it was time to retire. I held a press conference at City Hall to announce my retirement.

"I suppose, now that you are leaving the city, you'll probably be working for Con Ed?"

"Who, me?" was my reply. "In all likelihood, I'll be working against them!"

Sure enough, two days later I received a call from James Cope of Selvage and Lee, a public relations firm for the Scenic Hudson. This was an environmental group opposing Con Ed's 2,000,000 KW pumped storage plant proposed to be built at Storm King Mountain. Scenic Hudson was opposed because of the adverse effect on the environment and the damage to the fish in the Hudson. I was retained to provide technical testimony against the project.

I knew that I must come up with a constructive alternative. After considerable research, I learned that three successful installations of peaking gas turbines had been made by Public Service E&G of New Jersey, Cincinnati G&E, and the City of Holyoke. The operators were highly enthusiastic about the turbines' performances.

I presented my conclusions to the New York State Legislative Committee on Natural Resources and they unanimously agreed with me that the gas turbines would provide greater flexibility, resulting in lower costs, and that meeting the peak times would be more reliable with the elimination of transmission lines and the placement of units within the city.

A rehearing was requested before the Federal Power Commission, but they refused. It was only after an appeal to the U.S. Court of Appeals for the Second Circuit, who issued a landmark decision, criticizing the FPC for not searching out alternatives, that a rehearing was held.

Despite my presentations and that of other experts, the FPC examiner recommended building Storm King. After years of legal sparring, Con Ed finally, in 1980, agreed to abandon the project and installed 2,069 MW of gas turbines.

Garabed Hovhanesian

I was hired at GE (General Electric) as a design engineer. My interview had led me to believe I would work on motors, but that changed. I was told that because of my experience with metal, I was assigned to work on advertising, special tower and system programmable clocks.

My manager was an alumnus of a Technische Hochschule. His affected airs, as a wise German technocrat, were quite transparent. The rule of operation was to do model shop, conveyor and cook-book engineering.

My single contribution on this assignment, among many, was the design and implementation of a fusible line resistor to prolong the life of two 25 watt lamps which were in the clocks. By this means, I was able to stretch the life of lamps from one thousand hours average to eighty-seven hundred hours and, at the same time, provide a firesafe product.

There was no training as such. My manager resorted to browbeating tactics rather than professional leadership. I had my own motivation, I did not need his persuasion. I sought professional counsel from two fellow workers who were WPI alumni, Henry Deane and Jim Robertson. They were superb. Our manager of engineering was C. Herbert Ridgely from West Lynn. I finally asked Herb for a transfer to anywhere just to get away from the absurd management practiced by my immediate manager.

I was transferred to a branch plant as a resident engineer (1949). Here I literally flew; I studied quality control, machinery control and practiced my skills to the utmost. I learned and soaked up all the skills of manufacturing operations, relationships and productivity. I also learned something else, I found I had considerable skill as a teacher and counselor to plant personnel.

In 1952, I was asked to return to the main plant as a cost reduction engineer for the engineering department. Concurrently, I went to the Northeastern University Graduate School of Business where, in 1954, I earned my MBA. This was done with my own resources. The company did not see fit to pay for some of the good courses the school had. I could be wrong, but I attributed this to a sort of discrimination. During that period and succeeding years, I tried to write papers for AIEE publications but the legal red tape I had to go through to clear "proprietary" information was too discouraging.

I was assigned to an advanced development group to develop a new automobile clock. I docketed an idea for a constant torque motor to power an escapement device. Much to my surprise, my former manager got the patent described in my docket. When I apprised Mr. Terry of what had happened and proved it, he apologized and made a commitment to recognize my service at the next appropriate opportunity.

I applied for an opening in the General Electric Syracuse semiconductor operation. I found out that Mr. Terry had blocked an interview with the thought that: 1) I would be better off where I was and 2) that some things were opening up. I was also offered an opportunity in West Lynn and was likewise barred. Very soon thereafter, I was asked one day to appear at eleven a.m. in the mechanical and electrical test lab. As I walked in, the manager announced that I would be the supervisor.

I reduced the manpower, increased productivity and more importantly raised the level of engineering competence of the staff. I converted it from cut and try, to mathematical modeling, test planning and professional presentation. Cook-book engineering was gone. In 1959, after five years, I was asked to go into Advanced Development.

My task in Advanced Development Engineering was to design a conceptualized electric toothbrush (model shop sample), design and build five hundred close- to- final design samples for clinical testing by March, 1960 (from September l, 1959), and complete the design engineering for September, 1960 product introduction into the marketplace.

In the toothbrush project, I helped formulate ni-cad battery technology, 1.2 volt dc pm motor technology, charging circuitry, magnetic analysis of the induction charger, and designed and introduced into wide use the face gear (Fellows gear) for long life and fault tolerant assembly. We had achieved one thousand hours of continuous and intermittent life for motor bearings, brushes , commutators, gears, and plastic bearings (ABS), hermetic sealing of all live parts, and scotch yoke design to prevent stalling of the reciprocal motion (my patent) and more.

Beyond all this, however, I was instrumental in converting a venerable old clock factory into a multi-disciplined, multi-product operation. It became apparent to management that here was an opportunity to expand the product lines, hence a New Product Engineering Section was formed. But I was not appointed manager. A friend of a friend fresh out of Monogram, England by way of Asheboro, North Carolina, got the assignment.

The project to which I was assigned was to try to complete development of an electronic thermostat for fry pans, coffee makers, irons, toasters, hair dryers and so forth. The zero voltage switch had been developed but my predecessors both at Bridgeport Advanced Development Lab, New Britain, and elsewhere, had been unable to apply it correctly or at all. Within one month, I had designed the logic of the system and with my experience in putting products into production, I made a working prototype which performed admirably. The management at Bridgeport was ecstatic and asked me to demonstrate the technology to the engineering sections in Brockport, Allentown, Fort Smith and Asheboro.

I received personal recognition but my plans were for naught. It had already been decided to close the New Britain plant and centralize coffee maker production in Allentown. At about that time, I was offered a job in Allentown, as a sidewise move, without even an interview.

I refused on the grounds that I was ready, willing and able to be Manager of Engineering in Allentown. I was then appointed Manager of Engineering and Quality Control in New Britain effective immediately. This of course incurred the ire of the General Manager in charge of the plant. I stayed until the last working day.

In the morning of my last day, I was called by a friend John Macleod, Manager of Industrial Sales in Ashland, to meet with him for lunch at Bridgeport. There I met Bill Newing, Division General Manager of Finance, and Harry Stinson from marketing. Macleod asked me to report to him on Monday morning in Ashland where he would announce that I was to be his manager of industrial product planning. While I was doing that, I was to plan the Singapore plant.This was April, 1969.

My task was to covertly plan the transfer of the complete timer manufacturing operation from Ashland to Singapore. After the plan was completed and accepted by headquarters, it would be revealed to the local plant manager whose job it was to make the transition as easy as possible, which also meant heading up a public relations program. I would then be able to openly recruit personnel, set up a purchasing and delivery plan for all the equipment and write the financial, manufacturing, engineering, and manpower plans for the next five years with expansion plans embodied in the system.

Much later, I learned that corporate headquarters had promised Prime Minister Lee Kuan Yew that a vertically integrated appliance manufacturing plant would be built and operational by July, 1970. I suspect that the delay was due to being undecided who or what kind of person should be chosen to carry out this project. Early in 1970, I was named Project Manager. In September, I was named President and Managing Director of General Electric Housewares Pte. Ltd-Singapore.

In the spring of '71, I, with a management team, set up temporary training operations while the factory was being built. On July 1, 1971, as I had promised the company, we started production of timers. Our new factory had some two hundred and thirty thousand square feet with plans and room to expand. Under Bill Newing's management, I was given free rein to do the job. It was difficult learning the customs of a new culture and its government and a myriad of details associated with the implementation of the master plan. I loved every minute of my experience and never regretted going there. Mr. Borche, GE's CEO, comments were, "This is how a small appliance plant should be laid out. It's the best I have seen."

By June, 1972, the plant was fully operational. It became an attraction for the politicos of the company. In June of 1972, I was unceremoniously asked to return to the states. My replacement was a fishing buddy of Vern Cooper's who had risen above my manager, Bill Newing. I was also told that my transfer was no reflection of my work or accomplishments. I had an extremely good working relationship with Singaporean officials and the labor force in the plant.

When I returned to the U.S., I was given "corridor" assignments and stationed in the Ashland Plant. A mechanically operated digital clock timer was about to go into production. It contained no less than one hundred intricately molded and fabricated parts which required precision fabrication. I was under great pressure to see to it that the parts were made correctly. The product worked, but the accumulation of tolerances and the lack of precision processes to make the parts put me in the hot seat.

Concurrently, the plant manager was being assigned to Allentown and a new manager fresh from an Ireland plant was to take over. The domestic transferee was completely distracted by the prospect of his new job and gave all his attention to Allentown. The manager from Ireland spent all his time concerning himself with what was going on in Ireland and placing his re-patriates stateside. When , I surmised, headquarters had finally understood what was happening, they charged the new manager with some lack of performance. I do not know what. He then threatened me with firing if he was to be fired.

On February 25, 1975 I had a myocardial infarction. Incidentally with all of the turmoil, the plant grossed forty-one million dollars in sales, the largest in its history, and proportionately larger margins. This I attribute to John Haggerty's and my untiring efforts to keep things going. Shortly after my recovery and during my home recuperation, the manager from Ireland retired. A new quality manger was appointed, and when I returned I was made Manager of Manufacturing Administration.

Within a year I was asked to take retirement, I was fifty-five. During my discussion with the current plant manager, he told me that a number of people were after my scalp for no better reason than, "I was just to go." When I indicated that I would call Fairfield I was asked to hold off. The next day I received a call from John MacLaughlin. As a result, I was subsequently transferred to Bridgeport as Liaison Specialist, Overseas operations. All in all, GE did not treat returning expatriates too well or with much planning.

In 1983, I had bypass heart surgery; in '79 and '84, I had hernia surgery. In 1983-84, the division was sold to Black and Decker. I opted for a General Electric retirement at sixty-three.

Alfred Holtum

In the spring of 1954, the Signal Corps was setting up a Communications Department at the AEPG (Army Electronic Proving Ground) at Fort Huachuca, Arizona. They were looking for some adventurous engineers for key positions to form the basic structure of the new organization. I was interviewed by Ben Blom, the Department's new civilian chief, who offered me a branch chief position (in the Radio Branch of the Radio Division), which was scheduled to be a GS-13.

In April, the Signal Corps sent about a dozen key people to Fort Huachuca on TDY (temporary duty) to set up the department. Our first task was to write job descriptions of each of the positions planned to insure that they would support the grade levels proposed. Amid the arguments and hassles of the personnel experts from Fort Monmouth, New Jersey, whose goals in life were to reduce each proposed position by at least one grade level, I learned the significance of key words and phrases such as *Insure, Direct, Manage, Supervise, Under the Direct supervision of, perform, and other duties as assigned.* Using them properly is the mark of a true bureaucrat!

For the next three years, I would be involved in the technical administration of communication contracts, including the Army Division Grid System and other Signal Corps programs. I wrote the specifications and acted as the Contracting Officer's Technical Representative (COTR) for the first program to help establish tropospheric scatter as a viable communication tool for Signal Corps use.

In those days, fraud was minimal, at least in comparison to the transgressions perpetrated by contractors and government representatives today. There was a saying among government personnel, who administered contracts: "I've been insulted but never bribed." High level government officials were aware that good ethics came from the top down and made a conscious effort to remain above suspicion.

One of the chief concerns of the COTR was to insure that the contractor was fulfilling the specifications of the contract, and the government was getting full value for its money. This led to conflicts when a high ranking officer would request a demonstration for visiting dignitaries, the cost of which was beyond the scope of the contract. I got the feeling that some of the military believed the work being performed was chiefly for their amusement rather than the completion of a definite technical objective. When I would deny such requests, it caused consternation and enmity that was sometimes hard to live with.

One of the amusing bits of protocol at the AEPG back then was the practice of attaching one's title and job description to the front of one's desk. That way visitors could ration their approbation and respect accordingly. I recall one high ranking officer returning to apologize for his rudeness. "You looked so young for a position at that level," he explained.

When my boss, Ernie Stuermer, Chief of the Radio Communications Division, left to take a position in General Administration, I got his job at the GS-14 level. With that new job description pasted on my desk, I got more respect. I also discovered a trick that would get some attention. When I called someone and his secretary answered, ready to put me on hold, I'd say "This is A.G. Holtum." They thought they were talking to the Adjutant General.

I'll always remember one incident when a military personnel officer, a Lt. Colonel, came to review my job description. I later found out he was looking to replace some civilian employees with military personnel. "I've read your job description," he said, "and I think you'll agree that it's pretty general. What I'd like to have you do is make a long inclusive list of your specific duties, then if you should die or leave, someone else could step in and the department would not lose continuity."

I couldn't believe that an officer could be so ignorant. I waited, suppressing evil desires to lash out with vitriolic remarks (and many came to mind, none of which was flattering to the military). I told him that if my job were that highly structured and routine they wouldn't really need me and I'd be long gone. I further explained that every day was different and specific tasks unpredictable. I had the feeling he didn't really believe me, but I never saw him again.

With the remodeling of barracks into offices and the influx of more Signal Corps officers, the base was becoming more military oriented. More authority, mostly repressive, came from the top down and less innovation was coming from the bottom up. Of course, there were always a few bright officers, but they weren't the ones getting promoted. They were usually bucking some of the top brass. I'll always remember the words of the chief of the Department, a bird colonel, with whom I had a conversation several years after I had left the organization.

"I decided to back your program," he told me, "because I believed it had real merit. Besides, at that point in my career, I knew I wasn't going make general, anyway."

With the military personnel becoming more repressive and the long commuting distance tiring, I took a leave of absence for a year to take some graduate courses in administration and do some consulting. I went into business with Joe Boyer, the inventor of the V-37, a multiple frequency ham antenna that the AEPG was considering adapting for the MARS Program. We formed a little company specializing in antenna development. The adventure became short lived when we lost our financial backing.

One of the companies that had considered buying out our little business was the Andrew Corporation, headquartered in Chicago, Illlinois. When Dr. Andrew, its founder and physicist-turned-economist, saw our messy books he decided to just make job offers to the few key employees instead.

Meanwhile, I had been interviewed by several other companies including Hallicrafters in Chicago. While in Chicago, I stopped to visit the Andrew Engineering Laboratory located in a renovated barn in Orland Park. I was impressed by its head, Dr. Richard Yang, who some years later would become my advisor at IIT. Then after being wined and dined by the Director of Marketing, Robert P. Lamons, I was offered a position as Chief Engineer in the Andrew California Corporation.

My first task was to form a small engineering department. Two technicians, previously with our recently defunct antenna development company, were already working there. In addition, I hired two draftsmen and an electrical engineer, and was in the process of setting up a quality control department.

I have a penchant for walking into chaos and confusion. It happened with our first significant contract. The sales manager at Andrew California had already bid on a job retrofitting some trihelical antennas for shipboard application. This resulted in a small disaster when critical errors were found on the rotating mount drawings. While finishing up this project, I looked to the marketing department back in Chicago to recommend products that we could manufacture. They replied with some impatience and seemingly disinterest to "make anything that we don't make in Chicago; be your own marketing department."

I realized we would be pretty much on our own and decided, after talking with some of my cronies back in Fort Huachuca, to design and develop a series of telemetry antennas. This turned out to be a good market eventually, but our first sales were to my old friends at the AEPG. They bought one or two of each model. When they asked me how the new product was selling, I exaggerated and said, "Great."

"So, how come we got serial no. 1," the buyer said.

I learned never to start with serial no. 1.

I should point out that Dr. Andrew and most of his executives were always interested in educational potential. They were always financially helpful in furthering the education of their employees, as well as the children of their employees, and foreign students through the Aileen Andrew Foundation, set up for that purpose.

One of the most beneficial things I did for Andrew California Corporation was to recognize potential, despite a language barrier. I hired Geza Dienes, a Hungarian refugee, in about the summer of 1958. Creative and energetic, he would take my place when I left Andrew California and go on to become a vice president.

Jack Brown offered me his job as Chief of the Antenna Development Section. This was right down my alley and I was elated. It was July of 1962.

Now I had about fifteen mature high caliber engineers and technicians reporting to me including a good draftsman and a first class machinist. We had some worthwhile programs going when the word came back from Bancock that four sixty foot parabolic dishes, manufactured and shipped some months previously, were not working.

The sixty foot dishes were installed as part of a tropospheric scatter link between Bancock and Saigon (later the site of the Vietnam war). Using new patented Hublock connectors for the structural members, it was Andrew's hope to revolutionize the large antenna field. Using 20/20 hindsight, I believe now that is why Dick Yang left when he did. His section had furnished a good electrical design with proper dimensions and tolerances. However, the mechanical design section did not understand that the surface tolerance for the dish surfaces impelled total structural integrity with respect to the focal point of the paraboloid and really had to be met. Test reports were coming back that the system was not working. I inherited the problem.

Using formulas derived by my old friend, John Ruze, in his doctoral dissertation, we amassed curves and data to test a full size model constructed just outside the factory. When the physical dimensions were carefully checked, my curves projected a gain loss of more that 15 db.

When I reported this to Russell Cox, the CEO of the Andrew Chicago Corporation, in a casual conversation and told him the structures had to be replaced, he was appalled. He told others that I must have ice water in my veins to report a virtual loss of almost a million dollars with so little emotion. (That was a lot of money in those days, especially compared to my salary.)

The fiasco of the sixty foot dishes resulted in a reorganization of the engineering department. A young happy-go-lucky Ph.D., Ray Justice, was hired as Vice President of Engineering and section chiefs were raised in status to directors. A new Director of Mechanical Engineering, Dr. Keith McKee, formerly a professor at IIT, with a strong background in civil engineering was appointed. His newly designed replacement of the sixty foot diameter antenna looked strangely like a bridge. I became the Director of Antenna Design. With that new title, I was twice as smart!

There was a lot of good work produced in the next few years, including the axial mode helical telemetry antenna arrays used in the Apollo Program. One of my senior engineers, Larry Hansen, led the development effort to adapt our bi-filar helical configuration to this application. Our vice president cut the price to the point where we were almost sure to lose money, but he somehow felt the prestige of the job was worth it.

Kenneth G. McKay

A funny thing happened on the way to a nuclear reactor. In the early 1950s, many universities and industries believed that they had to own, or have access to, a nuclear reactor for research or for pedagogy. At the same time, the Air Force was enamored of the concept of a nuclear propelled bomber. They felt studies should be made of all the materials that might be exposed to neutrons in such a plane when most materials then become activated, i.e., radioactive, some more severely than others. A top-down decision was made that Bell Labs should have its own reactor that could be used for research and also for materials activation studies. The Air Force would contribute a substantial portion of the cost.

Having had a modest exposure to nuclear physics, I was given responsibility for organizing the project. I assembled a group of a half a dozen individuals from research and from our military operations. We proceeded to visit all of the AEC installations with reactors. Finally, we decided on a five megawatt copy of the MIT research reactor and plunged into planning the laboratory facilities that would be required: hot facilities, remote handling capabilities, glove boxes and other items unique to the manipulation of radioactive materials.

Meanwhile, I had asked one of our most ingenious experimenters, Walter Brown, to plan in detail a series of appropriate research experiments just as if we actually had the reactor up and running. He convinced me that it was much more time-consuming to perform an experiment in a nuclear environment than outside one; the cost effectiveness of a research reactor began to lose its appeal.

Coincidentally, the Air Force had second thoughts about their nuclear bomber and the materials activation program began to totter. This all came to a climax when the Air Force support was canceled and, as the project head, I did the unseemly thing of submitting a report stating that the last thing Bell Labs needed was a nuclear reactor. We would be much better off with a linear accelerator and, if needed, we could always negotiate for access to a port at the Brookhaven reactor. A week later the project was dead. I was transferred, with a promotion, to the Device Development Department.

In retrospect, we did the right things and came to the right conclusion, albeit with unintended help from the Air Force. It was, however, regretted that the Bell Labs president would never have his name on the Nuclear Research Laboratory.

Two years later (1959), as a result of some personnel shuffles, I found myself Vice President of Systems Engineering. Clearly, the term has more different meanings and interpretations than the term Research; I did a rapid study as to what it meant within Bell Labs. That too was shrouded in ambiguity. In brief, systems engineering was the activity of project definition and justification. In principle, a prospectus was prepared outlining the proposed system parameters and enough of a market forecast to justify proceeding with the development. In theory, all of the development engineers and their sponsors would be playing from the same sheet of music.

The major sources of conflict arose from the degree of detail contained in a prospectus. If the system was technically defined in too much detail, development engineers were prevented from exercising their own inventiveness, or else they simply ignored the prospectus. Conflicts arose when systems engineers tried to monitor the progress of a development too closely. Was the market justification simply for a go or no-go decision? Or was it the basis for a complete business plan? Some systems engineers so revered the prospectus that they regarded it as the true end product; it frequently would not be issued until the system itself was actually in the field. It was my task to define systems engineering at Bell Labs in such a way that these and other sources of conflict would vanish along with many turf battles, petty ambitions and other human foibles.

The development of all transmission and switching systems was headed by Vice President Walter A. MacNair. Walter was a stern task master but also a very thoughtful individual who was not enamored of systems engineering as it was then being performed. After some initial sparring, Walter and I agreed to begin discussions concerning the relative roles of our departments. We met alone twice a week, for about forty-five minutes, before the day's work began. Although sometimes interrupted, this procedure was followed for more than a year. No records were kept; we simply wanted a complete exchange of views and the exploration of as many possibilities as we could conjure up. We ended up in agreement on all major and most minor issues.

For example, the prospectus was retained but in minimal form; field trials were to be conducted by development engineers while systems engineers could participate. The transfer of a few top-notch development engineers to systems engineering would, in the long run, benefit both departments. We analyzed the gamut of interdepartmental interactions and emerged together.

Of course, the actual interactions took longer to resolve; a few had to await retirements. But things improved substantially and, of course, Walter and I become close friends. He even taught me how to shoot skeet.

Bellcomm. An interesting application of our systems technology occurred in 1962 when James Webb, director of NASA, appealed to the Chairman of AT&T, Frederick Kappel, for assistance in the Apollo Project. Bell Labs had established a fine reputation for systems engineering. It was felt that we could contribute effectively to the "Man on the Moon" venture, especially with its plethora of subcontractors. This was the genesis of Bellcomm, Incorporated. It was built around a core of systems members of Bell Labs' staff, which rapidly grew while assimilating the lore and science of space travel. Many significant decisions by NASA were subsequently based on studies by Bellcomm which helped bring order into a potentially chaotic situation.

The Board of Directors for Bellcomm was from AT&T Bell Labs, Western Electric and Bellcomm. In 1966, I became the Chairman and relished the opportunity of participating, tangentially, in the country's great space adventure. Actually, the real managerial work was done by Dr. Ian Ross, President of Bellcomm. He adroitly coped with the deployment of our forces, with the subcontractors and with the divisions and headquarters staff of NASA. It is not surprising that he later became President of Bell Labs.

As the successful end of the Apollo Project neared, I prepared to terminate Bellcomm. At first, the NASA director was taken aback—subcontractors never dissolved themselves. However, with his reluctant acceptance that our commitment had been fulfilled, we studied our options. The most attractive of many was simply to merge with Bell Labs, which was facilitated by the fact that we had identical pension plans. So ten years after Bellcomm's creation, we slipped back into the arms of Bell

Labs. In my own biased view, I believe that Bellcomm never received its rightful share of accolades. However, we had enough to go around and we were satisfied that a good job had been accomplished.

Military systems. The extension of my responsibilities to cover systems engineering for our military projects injected me into a wider world of endeavor. Our principal activities were in underwater warfare and the antiballistic missile project (Nike Zeus, later Sentinal and Safeguard). I was introduced to Kwajalein, that jewel in the Pacific, home for tropical suburbia (they had their own PTA). Here, drinking water was collected off the runway, and you could see the most advanced (at that time) radar in the world.

We demonstrated that we could hit a missile with a missile, but could we shoot down many missiles? As long as it cost more to shoot down a missile than it cost to launch another offensive missile, there was a problem. Every time we came up with another ingenious plan, Harold Brown, head of DDR&E, would simply double the concentration of offensive missiles. The ultimate response was: our GNP is greater than their GNP—we can outspend them. The project finally faded with echoes of the question: what do you do with eight thousand simultaneous offensive missiles?

The highlight of my military activity was the privilege of having Hendrick Bode as a member of my organization. Hendrick was a very well known mathematician who became intrigued with the analysis of complex military systems. Hendrick would rarely give a simple answer to an apparently simple question; he recognized that the apparent simplicity disguised inherent complexities. I can still picture him pacing back and forth in my office while responding to what I thought was a simple question. He would begin by discussing all of the peripheral issues as if they formed the outer edge of a spiral. He would then slowly circle inwards to the meatier parts and finally, after several hours, arrive at the center—a thoughtful and often unexpected answer to my query. His analyses were invaluable, but his approach infuriated some of the more action-oriented officers of AT&T.

AT&T. Bell Labs was a separately incorporated company with four shares of stock outstanding: two owned by AT&T and two owned by the Western Electric Co. which, in turn, was wholly owned by AT&T. In spite of such close financial coupling, a move from Bell Labs to AT&T was indeed an intercompany transfer. AT&T was a different milieu. In 1966, I was asked to become Vice President of Engineering at AT&T, a position traditionally referred to as "Chief Engineer." The AT&T chairman and the president urged me not only to take on the usual responsibilities but, more importantly, bring AT&T and Bell Labs closer together.

The formal ties were all in place but they sensed a lack of informal intercompany communication. My job was to fix that. However, I first had to assess my new surroundings. I had become accustomed to the deliberate easy going interactions between departments at Bell Labs. Now I found myself amongst a set of fiefdoms with limited intercommunications. I soon found that I should regularly eat in the officers' dining room: it was the best, and sometimes only way of finding out what was going on.

The volume of paper work was

substantial. If I did not understand why I should sign a document, I refused to sign. This caused considerable consternation but soon reduced the flow of paper across my desk. One of my first actions was to visit the library where I was regarded with obvious suspicion. It later emerged that officers did not go to the library; their requests came in by phone. Nobody browsed.

I adapted, and tackled my charter. The roles of AT&T Engineering and Bell Labs Systems Engineering approached each other and the line of demarcation became fuzzy. The communications improved—we shared joint responsibilities and that spilled over into the development areas. My main charge was accomplished with much help from others.

Meanwhile, I settled into my role as technical "guru" for AT&T although that term was never used. Twice yearly the Cabinet (AT&T officers) and the presidents of the nineteen operating companies would meet for a week-long conference. At those times, I was expected to give one or more speeches usually about the new evolving technologies and their potential impact on the telephone business.

This was a difficult audience. Although many of the presidents came from an engineering background, they had long since been enveloped by the press of operating problems and had delegated the care and fostering of technology to others. I discovered this on being requested for several copies of one talk, which rather pleased me, until I found that the presidents passed the copies down the line with the instruction, "Summarize this and tell me what McKay said."

So I visited all of the companies and attended many operating company conferences and Western Electric conferences and learned a great deal about the actual operating problems and successes of the business. I also learned to appreciate the capabilities and dedication of those who were brought up through the company ranks. It was very hard to find a bad apple in that barrel.

One day my career inadvertently took a new turn. An AT&T lawyer, George V. Cook, came into my office with a proposition: the FCC intended to hold hearings about vertical integration within the Bell System, and he wished me to appear as a technical witness for Bell Labs. This was to be a curtain raiser for the Justice Department's Antitrust Suit which would follow. George is very smart, thoughtful, and persuasive. I let my guard down and agreed. After all, it would be some way off in the future and would not take much time. I had already testified before commissions in New York and California so this should be no problem.

I prepared a lengthy submission detailing all of our arguments which was then redacted out of recognition by a platoon of lawyers. I was then simply to undergo cross-examination on that document. That is not quite what happened.

The FCC hearings are conducted in a court-like atmosphere under the aegis of an Acting Law Judge, an employee of the FCC. The attack (prosecution) is carried out by a team of lawyers, also FCC employees. (When one tires, another takes over). Unlike a normal judicial procedure, the rules are established by the judge and are subject to change without notice. By this time, I found that I was also representing Western Electric Company another source of many documents for me to absorb.

The actual hearings were exhausting. I

would be on the stand from ten a.m. to four p.m. The AT&T lawyer could intercede at any time; however, he chose to do so only when he thought I was in dire trouble. I suspect that he had too much faith in me because he rarely spoke up. At the end of the day, our lawyers would study the transcript to see what subsequently should be undone while I studied the new two foot high stack of documents from the FCC staff, some of which might be the basis for the next day's cross-examination. Their ploy did not endear them to me; we protested, of course, but the document dumping diminished only slightly.

Interest and tedium alternated during the hearings. The judge gave me substantial leeway in the length of my answers. I would follow several long answers with a one word answer. This would so upset the prosecuting lawyer that he would request a conference with his team. I wasn't trying to be cute; I simply wanted to throw the prosecutor off balance and I think that my several games did just that.

On my first appearance, I was on the stand for three weeks in a row. Later, I returned for redirect by the FCC lawyers and for some hard hitting cross by intervenors. I am still not convinced that legal-type proceedings are the best way of ascertaining technical facts. Since the intervenors attacked me, I attacked them. This left a residue of ill will which was not appropriate.

Bell Labs. In 1973, Executive Vice President Julius P. Molnar died. He was a powerful technical administrator at Bell Labs whom I admired and liked. His responsibilities covered all of the technical work at Bell Labs, except that of Research. I returned to Bell Labs to fill the vacancy. However, the scope of responsibility exceeded my grasp; Bell Labs had grown too large.

Just before the Korean affair, Bell Labs employed fifty-five hundred persons. Now the count was over twenty thousand and growing. It was decided to create two more executive vice president positions and divide the responsibilities into three parts. That greatly eased my load (I was still involved in FCC testimony) and enabled me to concentrate on what I liked best. I retired in 1980 before the divesture would force apart much of what I had helped to put together.

A. James Ebel

One of my consulting clients was the Peoria Broadcasting Company, which operated radio stations in Peoria and Tuscola, Illinois. In 1946 they asked me to come to work for them as their chief engineer. I really didn't want to leave the university because I liked the many engineering opportunities I had there. These included the approval of consulting, and the many perks one has as a member of an university staff. However, when they offered me double the salary I was receiving at the university, my wife and I decided we should accept the offer. (We had three children who were going to be attending college someday.) This turned out to be a good move, because I received many opportunities to participate in broadcast industry activities solely as the result of this move.

The company immediately had a number of projects for me. One was to design a directional antenna for station WMBD so that it could operate at 5000-watt full-time. With its existing non-directional antenna, it had to reduce power to 1000-watt at night. I had also convinced the management that FM was a coming thing and so an application was filed for FM. There was also the possibility of television in the far distant future, so consideration for the accommodation of television was to be part of the new transmitter location design. At the same time, the station in Tuscola, Illinois—a very small community—was to be moved to Decatur. Studies were to be made for a new frequency for full-time operation in Decatur.

The initial design for the directional antenna system in Peoria called for two 5/8 wave towers with nulls to protect other co-channel stations at night. One 5/8 wave tower would handle a FM antenna with a quarter wave transmission line isolation. The other would handle the television antenna, if television ever became practical. (That's what we actually thought in those days.)

Other co-channel stations objected to the two-tower design, as did the Engineering Department of the FCC at a hearing in Washington. Wilson Wearn, the Commission's Hearing Engineer at that time, cross-examined me critically because he felt that the two-tower Array would not provide proper skywave protection to co-channel stations. As the result of this critical questioning by Wilson Wearn for the FCC, the WMBD Legal Counsel Horace Lohnes, decided that we should withdraw the application and come back with a four-tower Array that the FCC would approve.

(Later Wilson Wearn as Chairman of the NBC Satellite Committee, became a member of the network Affiliates Satellite Committee, which I chaired. He also became President of NAB, supporting me financially in WARC [World Administrative Radio Conferences] activities, and a very close friend. I must admit that I had no idea of such a future friendship when he torpedoed my two-tower design.)

A four-tower Array was designed with one tall tower for daytime AM and FM, and television, and three additional quarter-wave towers. The Commission approved this design and a construction permit was issued. The Array was built along with a new transmitter building to house the 5-kw AM; and a high power FM transmitter. Again, when starting to tune up this Array, I ran into substantial difficulties.

Although there were phase meters and tower current meters in this tuning system—designed and built by Raytheon, the builder of the 5-kw transmitter, it was still impossible to get the nulls in the right direction with the correct field intensity. I assumed that the use of one tall 5/8 wave supporting tower, along with three quarter-wave self-supporting towers, was giving me a different current distribution than I had planned on in this design.

After working all night for over a month, I came up with a compromise pattern which I thought offered proper protection to the other co-channel stations, even though it looked different on a polar diagram than the calculated pattern. I took these measurements to Washington and went over them with Commission Engineers who weren't real happy with them, but grudgingly agreed to accept them—so the new transmitter location was licensed.

After the station had operated for approximately six months, the Commission called my manager and said there was something seriously wrong with this directional antenna system because it was not constructed according to the licensed design. The consulting engineer for a co-channel station, in working on a design for her client, noticed the error. My directional antenna design, in the form of a rectangular box, called for a spacing on the long side of the box of three hundred and forty-five feet. The plans submitted with the Proof of Performance showed a spacing of three hundred and fifty-four feet. Somehow (when the specifications were drawn up for construction) the forty-five had been inverted to become fifty-four, creating a nine foot error in spacing.

Needless to say, my manager was very upset with this development. After a number of conferences with the station's legal counsel, it was decided to hire the consulting firm of Lohnes and Culver to rectify the problem. I worked with them on measuring the current distribution on the 5/8 wave tower, actually doing the climbing myself. Ground current distributions were also measured. With this information, and using the tower spacing that was actually built, they came up with a pattern which would meet all requirements. The tuning for this pattern became very simple, and was quickly approved. I have to thank Charles Caley, Station Vice President and General Manager, for sticking with me in this "my darkest hour."

In 1946, WMBD filed for a full-power FM station. The other radio stations in Peoria also filed, so there were more applications than there were channels assigned to Peoria. Therefore it became necessary to have a hearing. The hearing was held in Peoria, and was hotly contested. WMBD did receive a grant for a FM station, eventually. In order to get a FM signal on the air, we applied to the FCC in 1947 for experimental operation with 250-watt (125 Vertical/125 Horizontal) to study the effect of polarization on reception. The principal to be determined was whether circular or horizontal polarization was best in the average community.

The FM station went on the air with full power late in 1947, initially with "elevator music" because AM and FM were not allowed to duplicate, and because the networks didn't want their programs on FM. Up to the time I left WMBD in 1954 to go to television station KOLN in Lincoln, Nebraska, under the same ownership, WMBD had not made a dime on their FM operations. A while back, when I came across this memorandum, I called the WMBD radio station manager who told me that FM now was far superior to AM—superior in quality and superior in earning capacity.

WMBD had a grant for Channel 8 television before the FCC went into a "TV freeze." Because of some uncertainty about the future of television, the station's legal counsel suggested that during the "freeze" WMBD turn in its construction permit since there would be no problem getting a permit after the "freeze" was lifted. This turned out to be a gigantic mistake. Because after the "freeze" was lifted, all of the radio stations in Peoria applied for Channel 8, which led to a long and bitter hearing.

I was put in charge of developing the hearing materials for our station. To prepare for this I spent a substantial amount of time visiting nearby television operations in Rock Island, Illinois and Chicago, Illinois. I also visited the Dumont factory and the RCA factory. In order to be better prepared, we purchased a complete set of studio equipment from Dumont; cameras, film chain, monitors, and so forth.

We demonstrated closed circuit television

in department stores in Peoria and a number of smaller communities. It was felt the ability we had gained this way would put us in good stead at the hearing (it didn't mean a thing). The book of exhibits showing the past operation of the radio station and the proposed future operation of the television station was over two inches thick. The hearing in Washington took over six weeks, and the record of the hearing testimony made a stack of books over three feet high.

The initial decision went to one of the other applicants who promised more in the way of programs than we had. We tried to be practical in the Peoria market. The winning applicant came up with proposals that exceeded what was being done in Chicago television, and the Commission bought it. After a number of appeals were filed, the Commission decided to make Peoria an all UHF market and gave everybody a station. Channel 8 went to the Moline-Davenport-Rock Island market.

While waiting for a decision from the FCC, Mr. Fetzer, a part owner of WMBD, purchased KOLN-TV in Lincoln, Nebraska. The Manager of KOLN-TV decided not to stay with the station. So on January 1, 1954, I was asked to go to Lincoln to manage on an interim basis, and to supervise as an engineer, the construction of a new television transmitter site. (When the station was purchased, the transmitter was at the studio site. The new transmitter site was to be twenty miles west of Lincoln.) Somehow the ownership never found a manager to replace me as Interim Manager, so I became General Manager and later Vice President and General Manager, and then President and General Manager.

Because Mr. Fetzer was a recognized television industry leader, and because of my past industry activities, I was appointed to the TASO Study Committee, the AMST Engineering Committee, and the NAB Engineering Committee. In 1964, I was elected a member of the CBS Affiliates Board. In 1968 when satellite interconnection between networks and their stations was initially proposed, the Board decided to set up a Satellite Study Committee so that the Affiliates would know as much about satellite use as the networks. As the only member of the Affiliates Board with an engineering degree, I was named Chairman of the new Satellite Committee, which I chaired continuously through 1989. When Satellite Committees were organized by NBC and ABC, I was named Chairman of the Combined Satellite Committees.

Our Satellite Committee immediately visited with Hughes, with NASA, with COMSAT, and every other source of satellite information we could put our hands on. When the delegation for the 1971 Space WARC was being set up, the Combined Networks Affiliates Associations arranged for my appointment to the delegation. I attended the Space WARC in 1971; the Broadcasting WARC in 1977; the General WARC in 1979; the North American Regional RARC in 1983; and the 1988 WARC on the use of the Geostationary Orbit.

My purpose at all of these international conferences was to represent the interests of the broadcasting industry—which included provision of spectrum allocations in the "C" band for station interconnection with their networks; provision for space in the Ku band for station interconnection with their networks; space in the Ku band for satellite newsgathering; and satellite remote broadcasting. We were also interested in protecting the concept of "localism" in the American system of broadcasting, which direct-to-home satellite broadcasting would not protect.

I became a member of the President's Frequency Management Advisory Council, representing broadcasting, in 1971, and have been a member continuously since that date. I was a member of the UHF Test Advisory Committee set up by the FCC; a member of the Satellite 2 Orbital Spacing Advisory Committee of the FCC; member of the NAB DBS Task Force; member of the NAB HDTV Task Force; and a member of a number of other sub-committees on HDTV.

Leo Berberich

My work at Mobil had attracted attention along with: publication of several technical papers; my activities in technical societies; the publication of my doctoral dissertation, and my association with Dean Whitehead at Johns Hopkins, resulted in a budding reputation in the field of insulating oils. As a result, I received an unsolicited offer to join the Westinghouse Research and Development Center in Pittsburgh, Pennsylvania. I was to head up a new group that would concentrate on the development of coil insulation for large power electric generators.

I was especially interested in this proposition because the work in Pittsburgh would offer not only the opportunity to work with insulating liquids, but also insulating solids and gases. In 1937, I accepted this offer and moved to Pittsburgh. For the next twelve years, I remained in the Westinghouse Labs. My responsibilities expanded to include work on insulation for other product lines and won me a promotion to the Manager of the Physical Chemical Section of the Insulation Department. I also served as consultant to the Westinghouse insulation user divisions and was often called to diagnose insulation problems in utility power plants.

In 1949, I made what was perhaps my most significant career decision. I was asked to leave the lab and venture into management to head what was then called the Liaison Engineering Department. This organization was a part of Central Engineering, reporting to the Corporate Engineering Vice President.

The department consisted of a group of highly trained specialists in various technical fields, including insulating materials, metallurgy, electronics, chemistry, ceramics. The department's main function was to assist the some sixty-eight operating divisions at Westinghouse in solving problems and keeping them informed of new developments at the Pittsburgh labs and throughout the company. My staff and I would have to travel to operating divisions located all over the U.S. and to the Canadian Westinghouse subsidiary located in Hamilton, Ontario.

With misgivings about leaving the lab, I accepted the position, but only on the condition that I could still act as a consultant in my chosen field of electrical insulation in addition to managing the operation. In a short time, our operation developed the reputation for knowing "who could do what both inside and outside the Westinghouse family."

Nine years later, in 1958, another opportunity arose. I was offered the position of

Manager of Engineering in the Westinghouse Electrical International Company, headquartered in New York. The offer was based on my broad knowledge of the Westinghouse headquarters and operating divisions in U.S. and the Westinghouse product lines. My staff, twenty product specialists, and I were responsible for servicing some one hundred and sixty-five product licensees located in North and South America, Europe, Asia, and Australia. The organization would supply engineering and manufacturing information to the licensees on all product lines covered by their respective license agreements. These product lines ran the gamut from nuclear power plants to refrigerators and light bulbs.

We also received requests for information from the licensees world-wide, obtained the information from the appropriate operating divisions in the U.S. and transmitted it back to the requester. In return for this privileged information, the licensees' agreements with Westinghouse required them to pay to Westinghouse a negotiated royalty fee ranging from one-half percent to five percent of the selling price, depending on the novelty and sophistication of the product line. I was also responsible for the ten resident licensee representatives whose sole function was to expedite the flow of information from Westinghouse to their respective companies. Since the expense of generating and servicing the licensees was nominal, a high percentage of the royalty income went right to bottom line.

I enjoyed this position since it enabled me to travel to almost all of the free world to consult with licensees on problems or complaints, to encourage them to expand their product lines, and occasionally to obtain information of interest to Westinghouse. Information developed by the licensees, on the product lines covered in these agreements, was passed back to Westinghouse on a royalty-free basis.

The most interesting of these trips was a trip to Japan where I toured about ten Mitsubishi plants located from Tokyo to Nagasaki. At each of these plants, I was cordially entertained by the General Manager. I was impressed with their capabilities in converting Westinghouse information into reliable products for their markets. I also admired their industriousness, their company loyalty, their food, and their culture. I am not surprised that Japanese products have achieved worldwide recognition for quality and reliability.

In 1961, after only three years in the position, I was offered another international relations position by the President of Westinghouse International. I was told that management wanted to create a new position based at the Westinghouse European Headquarters in Geneva, Switzerland. And I was judged the best qualified person to head this activity. The title was Director, Associated Companies Activities—Europe. The main objective of this position was to provide special service to some sixty-five licensees located in Europe. I would be responsible for an area from Oslo, Norway to Rome, Italy. In addition, two Westinghouse resident engineers at Siemens Companies headquarters in Erlangen and Munich, Germany as well as several young design and manufacturing engineers on training assignments would report to the activity. The assignment was to last four years.

I declined this offer, even though the challenge of the position, the concept of living in Geneva and the European travel sounded intriguing. In balance, the new position did not appear equal to what I already had in heading the Pittsburgh International Operations. However, three weeks after the first meeting on this subject, I was called back to the President's office and persuaded to accept this assignment. In August, 1961, my wife, Ida Mae, and our seventeen year old son and I sailed for Europe.

In Geneva, progress in developing the activity moved ahead rapidly, since I was familiar with the European licensees, having visited most of them because of my former job. The licensees in Europe truly seemed to appreciate the extra attention and service they received. I even had the pleasure of assisting and seeing my most important research and development project, the thermalastic power generator insulation, go into production in licensee plants in Oslo, Norway and in Madrid, Spain. My wife and I were also asked to represent Westinghouse at ceremonial programs, such as a ship launching by a steam turbine licensee in Holland. My youngest son was privileged to spend his senior year in high school at the Ecole International de Geneve, where the courses were taught in both French and English.

As an aside, it might be added that since the official language in Geneva is French, Westinghouse financed an intensive six week Berlitz course in French for us before we left New York. Although this was not sufficient to become fluent in the language, it was adequate to negotiate an apartment lease, shop, dine out and travel. Fortunately, most of the Genevois could understand and speak some English. More importantly, it was a requirement in the Westinghouse Geneva office for all secretaries to be capable of taking dictation and typing all correspondence and reports in English. All the licensees were already accustomed to receiving correspondence and information in English.

After less than three years in Geneva, I received discouraging news in the form of a letter from the same president of Westinghouse International Company who had strongly urged me to accept the assignment. The letter stated that, due to recent sharp decreases in Westinghouse earnings, management had reluctantly decided to eliminate this activity. This cutback would be part of the International Company's response to a corporate directive to reduce expenses.

I was assured that this was no reflection on my performance and unlike some others, I was assured of a position somewhere in Westinghouse. However, its location was not yet determined since the cutback was an overall corporate problem. After some preliminary investigations into opportunities based on my own knowledge of the company, my wife and I returned to New York during the summer of 1964, more than one year ahead of schedule.

Once back in the U. S., I spent several weeks investigating position possibilities. I concluded that the best opportunity for relocating was to the Defense and Electronic Systems Center near Baltimore. Baltimore management agreed that I should be able, based on my European and Japanese experience in particular, to assist in increasing the small amount of defense business Westinghouse had with our European and Japanese allies. Thus, in the fall of 1964, a new position, Manager,

International Programs, Aerospace Division, was created for me at the Baltimore Center.

I started out in the engineering department largely to become familiar with the product line. I later transferred to the marketing department, since my contributions to the international marketing aspects were potentially more important than the consulting aspects. I soon discovered that our allies' major interests, particularly the British and Japanese, were airborne radar and airborne counter-measures systems.

Since I was now back in the domestic part of the company, I worked with my former International Company colleagues first to obtain U.S. Air Force clearance to talk to our allies; then to arrange the necessary presentations to the interested foreign allies; and finally, after an order was negotiated, to obtain State Department clearance to export the products. In this activity, I became a frequent visitor to the U.S. Air Force at the Pentagon, to the State Department and to the appropriate foreign embassies in the Washington, DC. area.

The British Royal Air Force (RAF) was first to place an order. The RAF wanted several hundred Westinghouse radar systems for their American made F-4 Phantom aircraft. They planned to produce some of these systems in the United Kingdom (UK) after they obtained the necessary production license from Westinghouse and clearance from the State Department. Ferranti, Ltd., located in Edinborough, Scotland, was chosen to manufacture the system in the UK under Westinghouse license. I successfully facilitated Ferranti's production startup on this system. Some time later, the Japanese Air Force placed another significant order for the same F-4 Radar System.

Several allies displayed considerable interest in a counter-measure system developed for the U.S. Air Force by Westinghouse, but no orders were actually written until after I retired. During those six years plus before retiring, the proportion of international business increased from about five percent to more than thirty percent of the Aerospace Division's total business. I cannot take all the credit for this increase, there was tremendous cooperative team spirit exhibited among my colleagues at Baltimore and in the International Company. I do feel, however, that the international expertise I brought to Baltimore significantly benefited both my colleagues and Westinghouse.

Chapter nine

Retirement

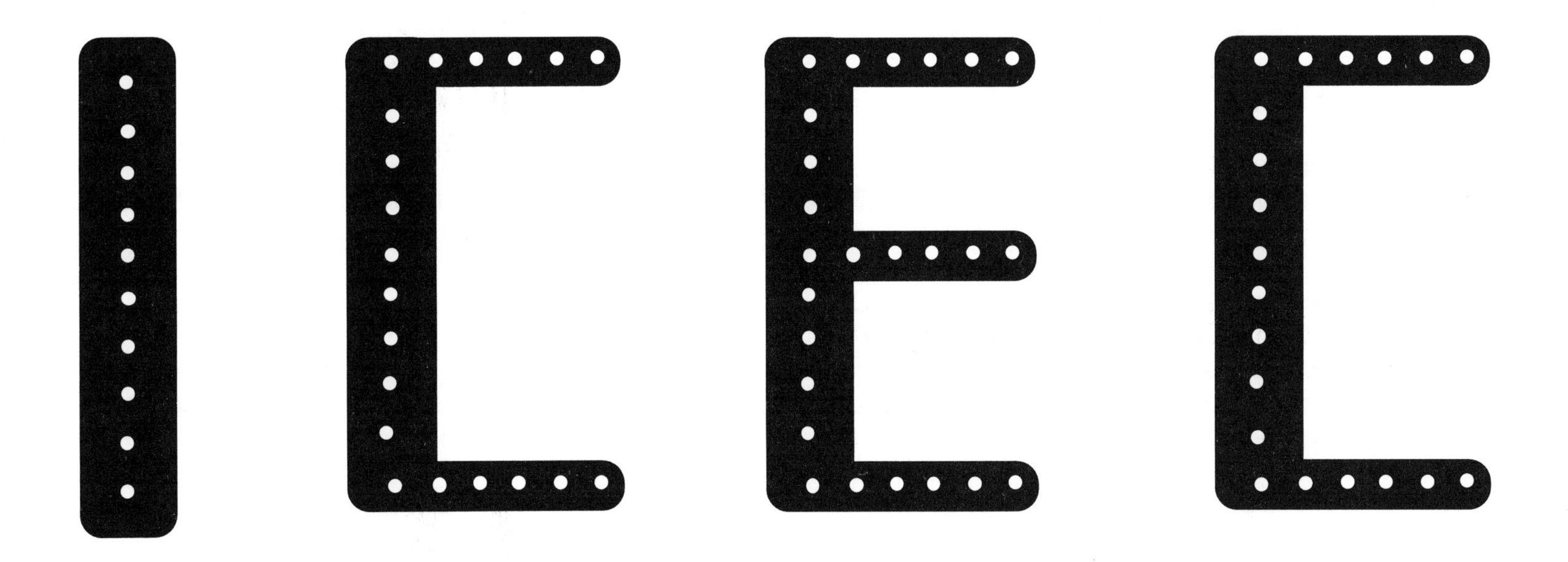

Cyril G. Veinott

Many years before it happened, I had started to think about my future retirement. There were many misgivings and fears. Would I have enough means to provide sustenance and a good quality of life for my wife and myself? I did not have any vested interest in a pension from Westinghouse. I had worked for my last employer, Reliance Electric, only 17 years, so could expect little pension from them. I might have to supplement my income in some way. Many after-retirement ideas flashed through my mind:

Rural small motel?
A professional career?
Make travelogues?
Become a professional tour guide?
Nomadic life in a trailer?

Getting ready for ?????? In order to prepare for mobility, whatever form it took, we quickly decided to sell our house and go to live in an apartment somewhere. Dee was willing, although neither of us had ever before lived in an apartment.

With still no definite plans yet taking shape, we approached my retirement with some misgivings, to say the least. Meanwhile, things went on as usual. NEMA friends gave me a retirement party in New Orleans. It happened on my last trip as a Reliance representative to a NEMA (National Electrical Manufacturers Association) meeting. Paul Bodine came down all the way from Chicago to attend the party and read a poem he had composed in my honor! No one ever did that to me before, or since. Later, friends and associates at Reliance in Cleveland, gave me my final retirement party. It, too, was a very moving experience.

A unexpected invitation to teach. Suddenly, before I had actually retired, there came an unexpected invitation to be a Guest Professor in Electrical Engineering at Laval University. Laval is a French-speaking University in Quebec City, Canada. This surprise invitation was the thrill of a lifetime for me! I had fallen in love with Quebec when I first had visited the city many years before. The love affair had deepened with one or two subsequent visits. Better yet, French was a language that I loved, having studied it four years in school. The French language had been a sort of a hobby with me.

But the idea of Quebec winters bothered me. I had grown up in Vermont, and well knew about cold winters. However, I just couldn't turn the Quebec challenge down flat, so I offered to go there for half a year, until Christmas. To my surprise and delight, they accepted me, even for half a year. That settled it!

George Platts

After 27 years in management, I decided, at age 60, to take the optional retirement being offered by my employer. I would concentrate my efforts entirely on volunteer work. I can frankly say that since making that decision, I have enjoyed the most challenging management career of my life.

Faced with managing volunteers who, of course, do not follow "orders," I have had to become a persuader. I have to convince volunteers that what I thought should be done was really their own idea. If I succeeded, my idea became their idea because they believed in it—were not merely "sold" on it. Of course, I did not score one hundred percent because I was dealing with people of different backgrounds and different beliefs. If people are given an "order," they don't believe in, they will not do their best. They must be convinced of its value and be enthusiastic to utilize their abilities.

As soon as I "retired," a former business associate asked me to serve on a chamber of commerce environmental control committee of which he was chairman. In less than a year, he accepted an offer to manage an overseas operation. He asked me to take on the chairmanship, and I agreed to do so. I appointed some new members to the committee, one a regional editor of the local newspaper, one an attorney, and one a retired physician.

We met monthly and, in three years, we were able to write a county environmental control ordinance that was enacted. In addition, a county environmental control director was appointed at our recommendation, an annual prize for the best science project was established as part of the science fair. The county school system also agreed to our suggestion that environmentally related aspects of a subject be taught in public schools.

Then I began to expand my retirement activities. I had a deep interest in the Easter Seal Society, which treats the handicapped with several kinds of therapy. We had the good fortune to have a local organization known as the Junior Service League start this service.

The League had formed the Volusia County Easter Seal Society for Crippled Children and Adults, with its own officers and board of directors. I served as chairman of the Easter Seal Annual Campaign for three years, as treasurer for three years, and as president for one year.

Later, I was elected by the Florida Easter Seal Society as a member of its board of directors. I served for one year as treasurer, followed by two years as president of this organization, which oversees all the local service societies in the state. I was also elected a delegate to the National Easter Seal Society convention for three years, representing the Florida Society. I continue as a member of the boards of both the local and state societies.

John F. Bell

My mandatory retirement from Zenith occurred in 1978. The old management had been replaced after a series of events that could hardly have been coincidental. A course on "professional management," free and on company time, was offered to the middle management and engineers. The First National Bank of Chicago lost one million dollars, and I have never heard of its being recovered.

A member of Zenith's board of directors who was also on the board of the First National Bank addressed the stockholders as follows as nearly as I can remember, "We just can't say enough about what a remarkable job the Zenith management team has done. The engineering management has produced a product that has made Zenith number one. The marketing management has done a marvelous job of selling the product, and the general management has guided the company to its present success.

"But the time has come to bring in a professional management team to use professional management methods to move Zenith into the future!"

The stockholders bought it and the new management team made Zenith just like all the rest of the controlled companies. It just wasn't Zenith anymore.

I lost all thirteen of my advanced development employees. I was given a small office in the exact center of the new Technical Center in Desplaines, Illinois. I was given no assignment, but I was a member of a new industry standards committee, and the committee elected me chairman. As chairman, I was a member of the international standards committee and made several trips to Europe to attend those meetings .

Writing those standards was my sole activity until retirement in 1978—a challenging task, but had I not been somewhat self sufficient, I would have suffered from rejection and loneliness. I was not part of any engineering effort and not involved with any engineering group. After notifying me of my termination, they had a job they wanted me to do and I was asked to stay. But I had already made plans and refused.

Kenneth Sturley

Increasing age reduces one's value as a consultant, and my wife (an Oxford graduate) and I (we have now completed 54, years of cooperation) have turned our attention to voluntary activities. We lecture on behalf of our National Trust, custodian of many famous country houses, our coast line and land areas like our Lake District. It is a joint lecture effort which seems to go down well with a wide variety of audiences and we recommend it as a recipe for "growing old gracefully."

Leonard Malling

I had decided to specialize in the infrared spectrum. I worked on several IR systems that dealt with reentry observations from a plane in the Pacific missile firing range. I became interested in complex frequency analysis. I wrote and presented several papers on my work.

I was due for retirement from MIT. Receiving adverse criticism for my work from the staff, I went to the laboratory director and complained that throughout my career this was the usual result of my work whenever I carved out new technical territories. However, as I was now retiring, I wished to have a clean slate. He smiled. I was given a substantial raise in pay and received apologies from the staff. I learned that attempts later to by-pass my IR designs ended in failure.

Alexander Lurkis

Because I was back in engineering after retiring in 1964, I organized a Consulting Engineering Company—Alexander Lurkis Associates, and later Alexander Lurkis, P.C. The American Museum of Natural History hired us to revamp their entire electrical distribution system, design many exhibits and floodlight the facade.

Roche and Dinkeloo, Architects, retained us to floodlight the facade and fountains of the

Metropolitan Museum of Art. Many other museums called on our firm to design their projects. The last design of the firm was lighting the Central Park Skating Rink (NYC). Many localities retained me as an expert in environmental cases for appearances before the Federal Power Commission and Public Service Commissions. Many cities asked us to study and design new lighting systems including: Cincinnati, Washington, DC, Garden City, Miami Beach, New Orleans and Harrisburg. We received an award from HUD (Department of Housing and Urban Development) and the American Iron and Steel Institute for the Cincinnati work.

In 1981, I terminated my design work and have concentrated on testifying in electrical negligence cases. I spend a lot of time gardening at my home in Holliswood, New York and writing.

I have written two novels and one play that have not been published. When I attended the 1987 NYU Writers' Conference, I was the oldest conferee. Since my wife is an artist, we spend a great deal of time collecting oriental art. I have very few idle moments, now in the winter of my life, because I know not much time is left.

Alfred Holtum

In August of 1974, shortly after my father died, I applied for a job as a Contract Consultant with the R&D Department of the CIA. (I had been trying to return to a government engineering job since before I left Andrew.) After the usual red tape of investigation and polygraph tests, I was hired on a two year renewable contract to serve as a technical officer at a GS-14 level. After three and a half years, I reached the age of sixty when they were offering early outs for people with over twenty years service.

This was just in time to take a job as Assistant Professor of Engineering Technology at the University of North Carolina at Charlotte in August, 1978. The prospect of teaching at a college level was why I had pursued the Ph.D. in the first place. I taught engineering technology and electrical engineering for six years.

But of all my exploits, I believe my most valuable contribution, besides raising six kids with my wife, Betty, is teaching the young engineers and technologists at UNCC. I enjoy being able to pass on some of the tricks, short cuts, and techniques of engineering and applied mathematics that one does not generally find in textbooks. It was also the six years that brought me the most enjoyment and satisfaction. I have no sage advice to aspiring engineers. Let them make their own mistakes! Even if I had some, who would listen?

Paul Burk

I took early retirement from Aerojet and went to work for Optical Radiation Corporation, Azusa, California. There I was responsible for all of the optical designs for their product line of Solar simulators and W circuit board (to 30 inch x 30 inch) photo resist exposure systems and projection exposure systems. ORC also manufactures vision care products and entered the IOL (Intra Ocular Lens) business after I obtained a U.S. patent for an aspheric IOL.

Soon after retiring from ORC, I was awarded an Oscar for technical achievement by the Academy of Motion Picture Arts and Sciences. This was for a design that was assembled by Apogee Studios Special Effects Division of a monochromatic high intensity blue screen projector, (March, 1985).

Since then, I have been involved as a consultant on optical and electro-optical devices. Currently I am into computers and working on some innovative vision care optical products and their testing methods. The results of one of my studies was published in the May 1988 issue of *Journal of Cataract and Refractive Surgery*.

Herbert Butler

In 1972, I decided that I had enough. My wife agreed. So I retired from government service and we returned to live in New Jersey. I tried providing consulting services for a couple of years and found it financially rewarding but otherwise not very satisfying. Being on the periphery of things instead of participating in the action is a poor substitute.

I went back to school and found it quite a challenge. It was possible, with motivation, to do better than I ever did as a young person. It was a thrill to be awarded a master's degree in teaching, with the honor of Magna Cum Laude in 1976. Teaching, however, was another story. I submitted proposals to a number of colleges, offering to teach a course in the applications of space technology, i.e., earth resources, conservation weather communications, etc. No takers, only doctorates need apply!

I then tried teaching math and science in the public high school and found it very troublesome. In several cases, I found it almost impossible. A private school proved to be a much better bet, although it did not pay as well. However, discipline and the learning environment were excellent. I had some top-notch classes in physics that made it worthwhile.

I finally gave that up, but still do some teaching, working with an illiterate young man as part of the literary program sponsored by Monmouth County. I also deliver Mobile Meals to shut-ins, garden, swim and exercise, help my wife and take care of the house, I am reasonably happy but not quite content!

John E. Duhl

Another change, the loss of the color TV division, put my job on the line. I was the last one in to Quality Control and Westinghouse had to keep the QC manager of that division to wind up all the contracts. I was kept for six months to train him on my job—a period that brought me to early retirement age. However, I wrote some resumes to see if I was employable.

All the resumes were responded to only one said, "not now, maybe later." I went through an interview session held by Singer with 139 others and was picked for a second session this time just four—and I was the lucky one. The assignment was to be their Qualification Assurance Representative. I worked for nearly six years at that and retired at 65. Since retirement, Facet Industries has called me in, once for a six month job to organize their quality documentation and the second time a one week photographic job.

Edgar Van Winkle

My retirement lasted until January, 1979 when I worked as a consultant for Bendix doing heat transfer analysis for a gyro on a rocket assembly until I had earned the maximum allowed by social security. To meet additional consulting demands, I formed the EMPAC Corp. to receive the funds. Then I went back to Conrac in May, 1979 on a consulting basis through October. The main clients of EMPAC were Conrac, Lear Siegler and Simmonds Precision. Computer programs were also written and sold. Three times I was called back to instruct new customers in the use and maintenance of the E2C Air Data computer.

(The main interests of the EMPAC Corp. at present are artificial intelligence and robotics.)

Donald Schover

Although my company had led me to believe that I could work for them as long as I wished, work was in scarce supply in 1983, and my salary was up there. They had sent me to a series of lectures on the major aspects of retirement: health, relocation, economics, life style change. The sessions were helpful.

Nevertheless, I felt betrayed and cast adrift for a time. My wife was retired. It was a double jolt for us both. My wife arranged a consultant corporation for me. I have done some control systems work and a great deal of volunteer software program design for a volunteer organization (Executive Service Corps) but that's another story. My big love now is personal computer software and the amazing things one can do with the available tools.

P.S.: What magic fatal disease causes the destruction of creativity and skill on one's sixty-fifth birthday?

Philip Sproul

In my sixties, BTL got out of the ABM business and a mass transfer to Bell System work took place. After resisting a transfer to another state, I was assigned to a software development department. Here I developed requirements for revisions of the system, from contacts with the AT&T and operating companies, estimated the economic applicability of the software to the entire Bell System, etc.

Although I retained my supervisory position, I was not allowed to manage a group, although my immediate supervision assigned one to me at one time. Apparently, my resistance to moving so near retirement (four and a half years) did not sit well with my vice president. Nevertheless, I received some of my best raises in this period and invented a variation of the system which was widely used.

I was ready for retirement. Following retirement, I worked part time in my own consulting business for five years. I am now relatively inactive. During my last four years, I felt some discrimination because of my age. However, it did not affect my economic well being, so I put up with it. I had some fine associates at all levels and would live my life over as an EE.

Rudolph Steiner

As to the "highs and lows," during the twenties and early thirties I attended college. I held interesting jobs during the forties, but they were somewhat disappointing because of the low remuneration During the fifties and sixties, the financial situation had improved nationwide, but specific professional "pros" and "cons" continued to exist, not always knowing what caused them. There was rarely an effect by other departments on that of my activities. The general policy of each association from the highest to the lowest company echelon seems to have been: Don't rock the boat. This atmosphere was often quite discouraging.

My greatest accomplishment was gaining the confidence of my various superiors. My worst failure was probably to remain with employers whose business was on the decline without the employees' knowledge, instead of changing jobs while the changing was good.

When retirement time arrived, I was definitely ready for it after having worked for a period of 39 years. Moreover, if one, such as myself, worked for an agency of the U.S. Government or for its subcontractors, contracts and employment can be created as well as terminated on a moment's notice. This happened more than once. Seeing thousands of employees walk out the door for good, one loses appetite for that kind of professional life, even though it never resulted, in my case, in periods of unemployment. There is just so much one can tolerate of such depressing situations.

Looking back, the most important people I have met in my career were the military officers of that Navy laboratory where I was employed from 1949 through 1955. They were highly educated professionals, each holding at least one academic degree, showed great personal interest in the employees' advancements and encouraged them in many ways. They earned as well as bestowed respect. Features I sorely missed in almost every other appointment.

I am definitely glad that I lived my life as an EE. I am pleased, if not proud, that I worked on projects such as the Polaris, Apollo and B1, to mention but a few.

In the course of my engineering life, I acquired 13 U.S., 3 British, 3 Austrian, 1 Belgian, 1 Italian and 1 French patents, mostly on electromagnetic devices and many of them assigned to the U.S. Government Navy Department. I am the author of numerous technical papers, many of them published in AIEE/IEEE organs and presented at AIEE/IEEE and other events.

Joe Zauchner

I did not enjoy being forced to retire at 65, but at that time it was IBM policy. It was probably a good idea. I found a worthwhile endeavor. I have been teaching electronics as a volunteer at a local high school for twelve years. I still get help from IBM, which is very valuable.

M. Lloyd Bond

Was I ready for retirement? I *asked* for it! Am I enjoying it? Yes, thanks to a reasonably satisfactory income (almost all of it NOT in pensions, because pensions were lost in moving from one location to another), and to reasonable health. Am I glad I was an EE....I mean, am? Not only glad, PROUD! Am I glad to be a Life member of IEEE? Naturally!

Yardley Beers

In retirement, I went further afield and obtained a BA degree in British history with an honors thesis based upon experience on an archaeological excavation on the Isle of Man. Retirement has turned out to be very different than I had expected. I had always viewed it as an opportunity to do something productive that was not necessarily feasible for those in working life.

Originally, I had planned to go back to teaching, preferably in Third World countries. To gain some knowledge of the situation, I managed to get myself appointed in 1974 to a US AID and NBS committee advising the governments of Ecuador and Bolivia on standards problems.

As far as the work on the committee was concerned, I found it very interesting, and I believe that I contributed at least my share to the effort. However, in regard to planning my future, the work was disappointing. I discovered that my specialized experience was much too sophisticated to be useful.

For example, the national standard for time in Ecuador was based upon one high quality quartz wrist watch. While there should have been three or five of them on hand for back-ups, its accuracy was more than adequate to serve the needs of the local industry. There was no need of a high accuracy atomic clock. I could have taught some routine physics courses, but others who had been dong it regularly had better immediate qualifications.

Nevertheless, I did make an application to the Peace Corps. There were several opportunities for teaching physics that were very attractive. I was informed that my application had survived the first review, but then I never was informed later that I had been rejected. At that time, jobs for physicists were scarce. I learned that many young graduates, who were unable to find regular positions, were joining the Peace Corps to gain more experience. They then would have a better chance of obtaining a position later. I decided, therefore, not to apply again.

The University of Colorado allows residents over 55 years of age to audit courses free of charge. On this plan, I audited courses in theoretical physics, archaeology, and English History for three and a half years.

At the same time, I applied for volunteer jobs in the state, county, and city governments, and was rejected. In some cases, I was told that, if I attended the open meetings of a committee for a couple of years to learn how it operates, I would have a much better chance for selection. As my 70th birthday approached in 1983, I decided time was running out: I could not afford to take time to attend meetings for a couple of years just in hopes that I might get selected.

Having a great interest in British History and archaeology, I entered the University of Colorado as a candidate for a B.A. degree in British History under what is called, "The Second Undergraduate Degree Program." I received the degree, Magna Cum Laude in 1986. Also, I was one of the recipients of a Jacob Van Eck prize for student excellence.

When I entered, I had an understanding with my advisor, that I would do a senior honors thesis based upon experience on some archaeological excavation in the British Isles. Through contact with a radio amateur, Jack Etherington, GD5UG, I made arrangements to work on the excavations at Peel Castle on the Isle of Man. This was one of the most delightful experiences of my life. Not only did it enhance my knowledge of British History, but it later led me to develop some interest in medieval art and architecture.

Finally my interest in "people" rather than "things" as well as in governments became developed. It is most unfortunate that I could not have had my history training before I was department chairman or a government science administrator. The training is valuable for developing techniques which are useful in other fields.

While I have suffered some frustrations and insults, and I have not made as great impact upon my colleagues and upon society as I have wished, my life has been a happy one. I am thankful for many blessings. I have done most of the things I have wanted to do, and I have had many pleasant experiences. Retirement has been especially happy.

Edgar C. Gentle, Jr.

Ready for retirement? Yes, I was ready for retirement with the Bell System when retirement arrived. No, I do not feel I was a victim of age discrimination.

Max W. Kuypers

Was I ready for retirement? No, it was unexpected. I had been hired by a company for a specific project which was for a multi-computer control complex for a pipeline in India. The project was expected to take about three years. I accepted because they said my wife would accompany me during the installation phase.

As the project developed, management changes occurred and budgets reviewed. It was decided that I should go over alone with periodic rotation for home leave. In view of the projected length of the installation phase, I could not accept this and was terminated. This way a replacement could be found who would accept these terms. Unfortuately, since I had known the manager who hired me for several years, I did not get the conditions of my hire in writing. At 65, I was out of work when Houston was still suffering from massive oil related unemployment.

If I could start over would I do thing differently? No, I don't think I would want to miss the travel and people I have met throughout most of the world. Even so, I still have a few countries to visit for the first time.

E. Byron Lindsey

In 1987, I earned the Amateur Ambassador Award from Advanced Electronic Applications, an electronics manufacturer, for my amateur radio efforts. Former U. S. Senator, Barry Goldwater got it the following year.

My participation in AMSAT (an acronym of the Amateur Satellite Corporation) is another source of pride and delight. Through it, I have been able to rub elbows with astronauts and converse with cosmonauts.

During a recent space symposium in Atlanta, for example, a support team and I were able to link the meeting hall's public address system to an orbiting Soviet MIR space station carrying ham radio operators and cosmonauts, Musa Manarov and Vladimir Titov.

The space travelers' English greeting to gathering was the culmination of months of planning: coordinating the timing of the space station's orbit over Atlanta with the symposium schedule—another accomplishment of communication for W4BIW.

Chapter ten

IEEE

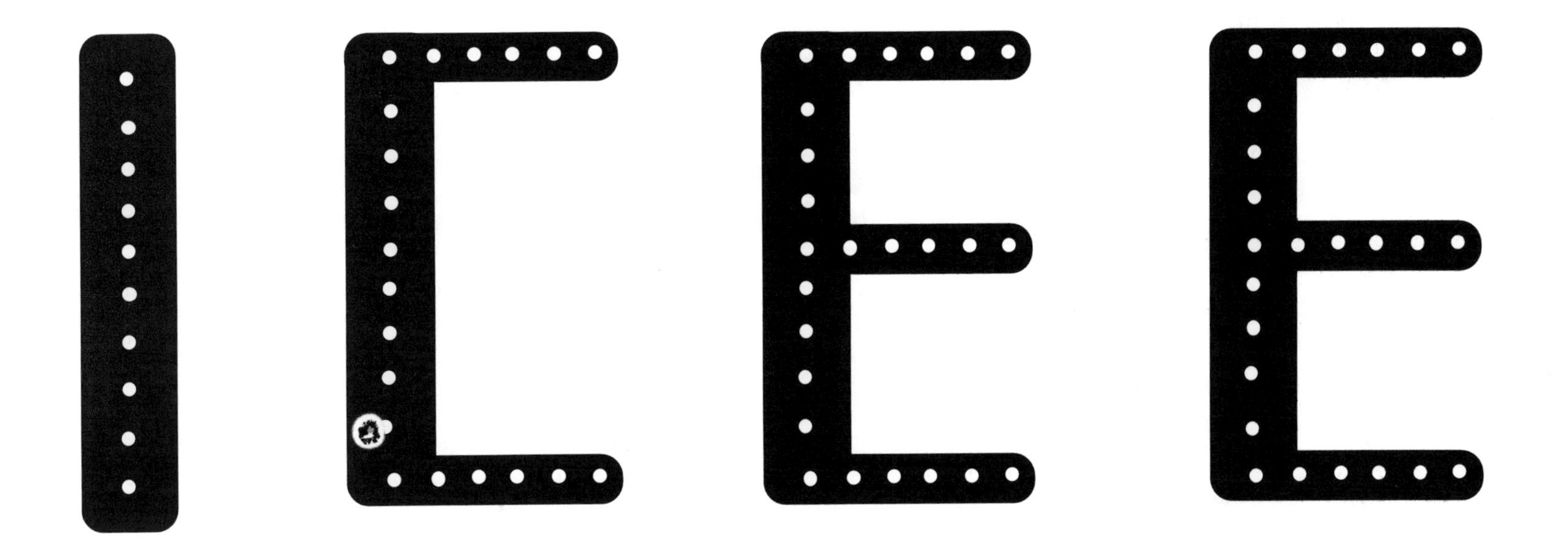

Harold W. Lord

I joined the AIEE in 1935. I became a Fellow member in 1951. In 1963, the AIEE and IRE merged to become the IEEE. When I retired from GE in 1966, I became an IEEE Life Fellow.

In about 1949, magnetic amplifiers had become such an important technical activity that the Electronics Committee of the AIEE (American Institute of Electrical Engineers) formed a Magnetic Amplifiers sub-committee. Dr. Edwin Harder of the Westinghouse Electric Company was the chairman. About two years later, the sub-committee had grown so fast that it was elevated to the status of a Technical Committee. I was appointed a member of it.

When the AIEE and the IRE merged in 1963 to form the IEEE, I was chairman of the Science and Electronics Division of AIEE and the Technical Vice President for Science and Electronics. This was 1962. I held that office for just one year instead of the normal two years, with the merger, the office was abolished.

After the merger, The Magnetic Amplifiers Committee became the Magnetics Group and the conference became the Magnetics Technical Conference. I was given the task of organizing an appropriate Technical Division and, after doing so, served as Division Chairman for a year.

The Magnetics Group rapidly grew became today's Magnetics Society. Under Dr. H. F. Storm, the Magnetics Conference became the present-day Intermag Conference.

For my "exceptional accomplishments and contributions to the field of magnetics," the Mag-netics Society awarded me their Achievement Award in 1984. That year I received the IEEE Centennial Medal.

Keats Pullen

I joined AIEE in about 1937. I was awarded the Caltech and Los Angeles area student paper prize in 1939. I had two letters published in *Electrical Engineering* in the early 1940s: one describing a matrix evaluation technique, and the other, pointing out the intermodulation consequences of the use of a field winding on a dynamic speaker as a filter choke. What I called "crossproducts" is now called "intermodulation."

In addition to these contributions, I have carried on independent research in a wide spectrum of electronics areas, ranging from circuits and active devices through propagation studies, and early development of procedures for calculation of patterns of complex antenna structures, including determination of the effective terminal resistance. Most of this work was rejected by various IEEE (and IRE) groups for presentation and/or publication.

I was perhaps the first to develop the theory of information engineering as a crucial element in electronics engineering, a subject not accepted in IEEE circles. I developed improved methods of presenting design data on active devices based on physical principles, another subject, likewise, not accepted.

I was the first to define the diffusion mode of operation of FET devices (overlooked by Dr. Shockley). This mode of operation was noted in a letter, *Proceedings,* IEEE, January, 1966. The basic principle of operation, disclosed in U.S. patent #3274462, is vital to computer MOS and CMOS applications and to the development of FETs capable of operating to 100 GHz and more. Many letters to editors have been published on this, but no papers were accepted.

Franklin Offner

I joined the Institute of Radio Engineers when I was a student. Then it was a technical and scientific society second to none. I opposed its union with the American Institute of Electrical Engineers. I believe it has resulted in the Institute becoming too industry oriented.

Robert McLane

I joined the Student Branch of the AIEE/IRE in 1946 at University of Illinois, Champaign-Urbana IL. In 1968, I became an IEEE Senior member. The organization has lived up to my expectations through the years, but I must admit, the *Proceedings* of the IRE were a bit "over my head." I found *Spectrum* most welcomed!!!

Vernon McFarlin

At the AIEE Winter Power meeting in 1959, I presented Conference Paper CP58-1310, "Tests and Life Expectancy of Generator Windings." This paper was listed among other references in a paper by a British engineer. I received a copy of the paper from an associate who maintained hydro generators for the Aluminum Company of Canada. It described the only method I had ever heard of for using test results to predict the life of electrical insulation. Some tests the author used were more sophisticated than Boston Edison's, but I still used his system of evaluation on the tests we routinely made on our generators.

On Friday, September 20, 1963 (the day I was fifty-four years old), I inspected one of the Edison Company's large generators and recommended removal of the field so I could make a more thorough inspection. Although this would involve spending probably well over ten thousand dollars, my recommendation was accepted without question.

After the field was removed, I was able to inspect the armature winding. It showed some signs of deterioration. The manufacturer's engineer inspected it and said to me, "Mac, that generator is thirty-five years old and looks it. It should be rewound."

I said, "We'll give it a high potential test and if it passes, we'll put it back to work."

The engineer responded, "You're taking an awful chance." It stood the test. The microampere leakage readings indicated a winding in better than normal condition for a machine that old and I told my bosses to put it back together. A rewind would have cost over one hundred thousand dollars.

At the Electrical Insulation Conference in New York in September, 1965, I presented IEEE paper 32C3-90 entitled, "Evaluating the Condition of Generator Stator Insulation by Test." I presented an updated report based on more data to the Doble Client Conference in 1973. An engineer from Ontario Hydro telephoned me after the Doble Conference presentation. He said it was the only reference he had found which showed how test results could be used to predict the life of electrical insulation. He was interested in how it might be applied to his company's generators.

J. A. M. Lynch

When I moved to Halifax, my IEEE membership was transferred automatically to the Montreal Section, which owned everything in Canada from the Ontario border to the Continental Shelf. Fellow naval electrical engineers had frequently remarked on the anomaly—the Maritime Provinces and Newfoundland had little in common with highly industrialized metropolitan Montreal and would probably be better off having their own IEEE Section.

But no one had done anything about it; certainly naval engineers had no time during a normal two-year appointment to do other than suggest. I decided it was worth a try and that I could at least explore the matter.

I wrote to that wonderful Emily Sirjane, explained my idea and asked for a list of members in the four eastern Canadian provinces—Newfoundland, Prince Edward Island, New Brunswick and Nova Scotia—and for chapter and verse about petitions to form new sections. I proposed the name "Canadian Atlantic Section"—the usage "Maritime Provinces" has never connoted the inclusion of Newfoundland, to Newfoundlanders or to "Maritimers."

Emily sent the materials: there were, as I remember, some one hundred and five IEEE members, widely dispersed. The requirements for forming a new section were very explicit and very restrictive; there seemed no way for any significant progress to be made in a mere six months. The most difficult hurdle was posed by the required format for the petition: Signatures had to be on one page, multiple pages of signatures had to be arranged in a certain order, that sort of thing.

I wrote IEEE headquarters requesting waivers giving reasons. In the case of the petition, I recommended I conduct a survey and that, if the favorable ratio exceeded the stipulated minimum to carry the proposition, the results be accepted in lieu of a formal petition. I drafted a postcard-sized format; it explained the proposal and included, more or less, the words of a formal petition, followed by tick-off boxes: "Yes" and "No." My office address appeared on the reverse.

Emily apparently put the package together in the proper form and carried it to the Board of Directors. The Board very quickly approved all the waivers. Emily immediately produced and mailed the cards directly to one hundred and five individual members in the four provinces. Before many weeks, some eighty-three of them arrived on my desk! What pollster or survey-conductor can match that?

I kept Montreal informed of my intentions and of my progress and dared to suggest that they might donate a couple of hundred dollars to the new section. They had been collecting per-capita returns from headquarters for members east of Quebec for a good many years. On a duty trip to Montreal, I had a layover and notified the Montreal Section of my presence in town. I had never met their chairman, but he gave me a cordial response—scooped me up, wined and dined me and promised me two hundred bucks—which was eminently satisfactory.

When I had the somewhat more than eighty returns from my survey, I seem to recall that I received one "No" from a member in Newfoundland, with a note that he enjoyed the Montreal newsletter. Two co-workers in a New Brunswick city voted "No," appending critical remarks to the effect that identification with Halifax was a far, far, worse fate than affiliation with Montreal. (I believe now there is an independent spin-off section in their fair city—I rest that case.)

When the score was eighty "Ayes," three "Nays" and only twenty-two "No Shows." I packaged them together, with a covering letter to headquarters, saying in effect:

> "It is not unknown for a sailor to sire a child, then sail away before the child is delivered conveniently avoiding all further responsibilities.
>
> "I am being transferred to Quebec City about mid-April and have handed over all correspondence on the formation on the Canadian Atlantic Section to Lieutenant-Commander Wesley D. Hutcheson, MIEEE. Wes has agreed to see the project to completion.
>
> "As I shall be rather busier at the job I'm paid for in Quebec than I have been in Halifax, I believe I can promise you that I will not pinch the Quebec membership from the Montreal Section next year."

The Canadian Atlantic Section was formed on 20 July 1966, with Wes Hutcheson as its first chairman.

In the autumn of 1968, in conjunction with a professional group convention, the Region 7 Annual Meeting was held in Montreal. I attended. In the lobby of the convention hall there was a booth labeled "Membership Information"; I did have a question and singled out a particularly lovely lady to seek an answer. Promptly and most efficiently, she responded.

Thinking I might help the Montreal Section by handing a deserved compliment to one of their volunteers, I said, "You seem to know what you're talking about. What's your name, please?"

"Emily Sirjane."

Now, *that* was a privilege. Indeed, we sat together during luncheon. I shall never forget her.

Winthrop Leeds

In an interesting quirk of fate that my life was affected in three important ways by a man whom I never met or even saw. This was Benjamin Garver Lamme, chief engineer of the Westinghouse Electric Corporation, who was responsible in his earlier years for the design of the first electric generators installed at Niagara Falls.

In my junior year in high school when I was deciding which college I should attend, my attention was drawn to the remarks of Mr. Lamme published in the September, 1921 issue of the *Journal of the AIEE*. His remarks were in response to a questionnaire submitted to one hundred prominent electrical engineers asking for suggestions to improve the college trained engineer.

He wrote, "As an indication of how little we care for the student's knowledge as a whole, I may say that in reviewing the selected college graduates for our work here, we ask them almost nothing about their work in general. We simply endeavor to find out their aptitudes, characteristics, analytical abilities, use of mathematics and how well they can use their heads." This confirmed my predisposition to attend a small liberal arts college as a foundation, with possibly a year of graduate study in a large institution to bolster my technical knowledge.

In 1937, I was awarded the Lamme Scholarship. This was given by Westinghouse for a year of graduate study to a promising young engineer at a school of his choice.

In 1971, a year after my retirement from Westinghouse following a forty-four year career in one field, the design and testing of high voltage power circuit breakers, I was presented with one of the IEEE prestigious awards, the Benjamin Garver Lamme Gold Medal. The citation read,

"For contributions to the development of high voltage high power circuit breakers, specifically using SF_6 gas, and for his effective exposition of the theory of arc interruption."

During my career, I advanced through the membership grades, becoming eventually a Fellow and Life member in IEEE. I prepared seventeen transactions papers, including a national prize paper in 1943 with L. R. Ludwig, co-author. My only national office assignment was Chairmanship of the Research Committee, 1958-1960.

Alfred Holtum

The IEEE (and the IRE) was of significant importance to my professional career, particularly in the way it enabled me to keep abreast of the state of the art. I have been a member of the Professional Group of Antennas and Propagation since its inception. I served in various offices of a local chapter while living on the west coast. It also enabled me to effect and maintain important contacts from a business as well as a technical standpoint. My affiliation served as an important sales vehicle to the Andrew Corporation and I helped "man their booth" in the annual New York City show as well as Wescon for many years. (One of my daughters complained that I could never be home on her birthday, March 26, because of the IEEE show in New York City.)

Edgar C. Gentle, Jr.

I joined IEEE (it was called AIEE) when I was in my junior year at Auburn; this was in 1940. Of course, when I was off in the military on active duty, I certainly wasn't active.

After WWII, around 1950, I became active in IEEE again. The people whom I met through the Institute did influence my personal and professional life. I recommend IEEE and similar organizations for engineers at any age, especially young engineers just starting out.

Norman A. Bleshman

In 1953, I faced my first professional organizational challenge serving as AIEE Student Branch chairman. I was responsible for organizing the first joint Student Branch activities among the three engineering schools in the area.

After graduation in 1947, I relocated to the New York City area where I soon joined a very active Power and Industrial Division, of the New York Section of AIEE. It is significant to remember that, unlike today, members worked competitively to achieve leadership roles in the Power and Industrial Division with hopes that some day one would become a Section officer.

To eventually achieve the position of Chairman of the Division, I labored long in the Education Committee (which I eventually chaired). During that time, I helped produce a large number of well attended educational programs focusing on the needs of the Power and Industrial Engineer These included courses that prepared engineers to pass the Professional Engineering licensing examinations.

We also held the first courses in atomic energy. In about four separate sessions, we served almost one thousand engineers. A by-product of these courses was the funds they raised which allowed the Division to afford other programs and activities for the members. In the 1950s, the Division donated, I believe, about ten thousand dollars toward the construction of the Engineering Society Building. This was probably one of the largest donations of an arm of the AIEE or any of the Founder Societies.

Gustav "Gus" Henry Bliesner

I was appointed to the position of Continuing Education Coordinator by IEEE National Manager, Vincent "Vince" J. Gardina, May 23, 1976. Vince's office was in the Service Center, 445 Hoes Lane, Piscataway, New Jersey. Portland Section's educational chairman, IEEE Fellow Stig Annestrand, selected me as the Sections Coordinator. Stig served as Manager of BPA's branch of Laboratories, Vancouver, Washington, from 1972 through 1980. Our various lecturers included Messrs. Blackburn and Buckley and Dr. Eccles in 1977-79.

Joseph E. Guidry

My membership in the AIEE, and subsequently the IEEE, has always been highly prized and held in high esteem. It has benefitted me greatly in enhancing professionalism, making many engineering friends, keeping up with the electrical profession and providing much social pleasure (to both me and my late dear wife, Juliette). The dates of my AIEE membership and my activities in the Washington, DC, Section were as follows:

1931 - 1933: AIEE Student Member while studying for my graduate degree at Tulane University.

Nov. 23, 1949: Secretary H.H. Henline notified me that the Board of Directors had elected me into the AIEE at the grade of Member.

1950 - 1952: Member, Technical Programs Committee Washington Section.

1952 - 1953: Chairman, Subcommittee on Power Technical Programs Committee.

1953 - 1954: Chairman, Technical Programs Committee.

1955 - 1956: Chairman, Membership Committee.

1956 - 1957: Alternate Delegate to the D.C. Council of Engineering and Architectural Societies.

1957 - 1958: Chairman, Student Activities.

1958 - 1959: Chairman, Attendance Committee.

1959 - 1960: Member, Awards Committee.

1960 - 1960: No official duties, Washington Section.

1961 - 1962: No official duties, Washington Section.

1962 - 1963: Harold W. Kelley, Chairman, Washington Section, stated that it was with mixed emotion that the 1962-63 Yearbook was being published. He noted that April 9, 1963, would be the sixtieth year of the Washington

Section and that the 1962-63 Yearbook would be the last published by the Section. On January 1, 1963, the new Society of the IEEE was created by the merger of AIEE and IRE.

Lester E. Haining

I first joined the AIEE as student engineer while in college and then the IRE when I started with RCA. I joined the IEEE when the AIEE and IRE merged. That association has been very valuable to me, and still is.

C. Richard Ellis

Having joined the IRE as a Student member, I responded to an ad in the *Proceedings* by General Electric for their new Electronics Park in Syracuse, New York. I was hired as a technician for the summer.

John J. Di Nucci

As I recall, I joined the AIEE during the first semester of my junior year in the fall of 1941. This student membership continued after graduation for the duration of my service in the Navy in WWII, with a moratorium on payment of dues, which was the Institute's policy at the time. My reason for joining was to avail myself to the technical papers appearing in the Institute's monthly magazine.

Melrose M. Jesurun

I joined IEEE (AIEE/IRE) as a student to keep abreast of modern technology. I enjoyed my life as an electrical engineer.

Robert G. Johnson

I feel that the IEEE organization has been a very effective one during my career and has been instrumental in advancing our technology into the computer age. I have enjoyed my engineering associations within and outside IEEE even more, I think, than my associations as a member of the American Physical Society, as a physicist. In addition, I feel that the IEEE insurance program has been a major benefit to the members. I have noticed over the years that many other societies have developed similiar group programs.

Fred Tischer

I liked teaching, particularly when talking about my research and about material developed earlier and now reformulated for teaching. The lectures, frequently interspersed with demonstrations, appeared to be well liked. The students at Ohio State University were usually highly motivated and appreciative.

After having received "the great surprise of my life" in 1962, the Fellow award of the IEEE, I learned afterwards that my students had made a drive to get me the award. I also learned that they were joined by one member of the faculty in their efforts. However, since I knew I was not a "favored son" of the majority of the department's senior faculty and was probably destined to remain an associate professor the rest of my life, I wasn't in a mood to settle down permanently.

When NASA offered me a temporary assignment as an "expert in space communications" in early 1962, I gladly accepted. Symptomatically, the request for sabbatical leave was rejected, so I left on a "leave of absence without pay" arranged by the dean's office.

Donald Schover

I often had the feeling that a lot of the very esoteric articles were written by professors and graduate students to impress each other and snow the illiterates out there with mathematical symbols and arcane formulae. All of my self-study knowledge gains have come from well written textbooks or good tutorial articles in technical magazines. The direct help I've received from IEEE is a couple of home study courses and the group insurance plan.

Theodore Schroeder

I served on the AIEE System Engineering Committee, 1953-58. I was elected to the grade of Fellow for "contributions to planning and design of power systems." I coordinated the campaign within the Central Illinois Section membership to financially support the construction of the new headquarters building at 345 East 47th St., New York City.

R. H. Eberstadt

One of the highlights of my affiliation has to do with the AIEE International Convention held in Mexico City during 1948. It was the largest convention that Mexico had ever hosted to that date. It was so colossal, in fact, the chairman had to take a year's leave of absence from his job to organize it.

A group of us had to form the organizations to render professional services (busses, translators, etc.) that were non-existent and needed for the convention. Personally, I seem to recall that I was Registration Chairman; I also had the responsibility to make sure that the hotel—which was going to be our "headquarters"—would be finished on time. There were a number of "near misses," but in the end, everything seemed to come off without a hitch.

Harold A. Wheeler

I joined IRE in 1927 (at age twenty-four) as soon as I became aware of it. Professor Hazeltine was one of the leading members of IRE at the time and our Chief Engineer was a member. However, they did not urge me to join as soon as I should have. In college, my professors had leaned towards the American Physical Society since I was involved in the physics curriculum.

Later I joined the older society, AIEE (American Institute of Electrical Engineers, established 1884) which in 1964 was merged with IRE to form IEEE.

Before World War II, the IRE grew at a healthy but not spectacular rate. It was fun. I attended most of the conventions and was acquainted with most of the leaders. Then, all topics were basically covered in *Proceedings*.

After the war, all that changed. The explosive growth led to more reliance on local sections. Publications had to be separated by Professional Groups (now Societies). Then the merger with AIEE further enlarged the bureaucracy.

Big is not beautiful. My local Section

(Santa Barbara, CA) has nearly one thousand members but the attendance at a typical Section meeting is only thirty (three percent). In the United States, I suspect that only one half of the eligible engineering graduates are joining IEEE. With all the Society Transactions, there is insufficient space for all the papers submit, even after condensed. I query whether the rate of growth of IEEE can continue, or even whether I want it to. Mostly I like its activities but I feel "lost in the shuffle."

Paul D. Andrews

In 1923 I became a member of the Institute of Radio Engineers. That membership was raised to the Senior level in 1943. I was issued a Professional Engineering license for New York State by the University of New York. I also became a member of the American Institute of Electrical Engineers in 1940 and am currently a Senior Life member of the Institute of Electrical and Electronic Engineers.

Thomas M. Austin

I joined the AIEE while a student. Also, while a student, I joined Tau Beta Pi, Eta Kappa Nu, Sigma Yi, and Phi Kappa Phi as rewards for scholastic excellence. I don't think I met any people strictly as a result of being an IEEE member.

Wieslaw Barwicz

I proposed to the government to organize the Industrial Institute of Electronics. My proposition was accepted. I organized the Institute with five branch establishments in different towns. Two of them some years later were operated independently.

George H. Barnes

I joined the IRE in my analog computer days, the early 50s. It has been an essential resource for just about everything I have done.

Aubrey G. Caplan

I joined IEEE in 1943 as a Student member. I am also a member of IES, CSI and ACEC. I served as the regional Vice President of the IES in 1976 and as local president of ACEC of the Greater Pittsburgh Chapter in 1980. Also, I was honored as "Electrical Man of the Year" in the Pittsburgh Hall of Fame by the Electric League of Western Pennsylvania in 1986.

Sidney Bertram

I became an Associate of the IRE in 1936 while with the Radio Institute of California. I have not been active in either the IRE or IEEE except as a contributor to the publications.

William A. Edson

I joined the AIEE as a senior at the University of Kansas under pressure from Dean Shaad and continued as a member for about fifteen years. By the end of that time, my interests were widely divergent from those of the AIEE and I resigned to join the IRE in 1941. I was Chairman of the Atlanta Section, 1948-49, and of the San Francisco Section, 1963-64 (during the transition to IEEE). It was through these activities that I met Bill Hewlett and Fred Terman, and eventually became Director of WESCON. These contacts had very beneficial effects on my career and personal life.

Glydus Gregory

I was first invited to join the IRE in the late forties or early fifties and I attended a meeting where Dan Noble gave a lecture. I did not join but I continued to attend some of the meetings and lectures, including the IEEE Area 6 Convention in Sacramento in 1957. There I was impressed by the youth and knowledge of the speakers. I joined IEEE in the late 60s or early 70s, shortly before attending another Area 6 Convention in Sacramento.

Harry D. Young

1947—I was attending the CUNY school of engineering and joined the Institute of Radio Engineers (IRE later becoming the IEEE) as a Student member. I subscribed to the IRE *Transactions* and recall a report from a mathematician at the Thomas J. Watson Mathematics Lab in Yonkers, New York—a research arm of IBM.

He decried an effort to analyze a rather complicated differential equation. He used IBM equipment consisting of multiplying punches and accumulators and went through several tens of thousand of punched cards.

The multiplying punch would read factors from fields in the card and after multiplying them, punch the result in a third field. This machine was connected to an accumulator that could sum the results and punch the sums in another series of cards. After describing the several months effort in detail, he concluded with the statement, "No longer is it a question that the problem is too complex to solve. Rather the question is, 'Is it sufficiently complex to warrant the effort required to solve it?'"

1953— Seyour Shefter was a mathematician in the Meteorology Branch at Evan Signal Laboratory. He had been going to computer conferences to see how the technology could be used to improve the processing of radiosonde tracking data. He would describe the events at the usual computer conference.

Typically, an engineer from one computer company would state their computer processed at the rate of seventy-five thousand operations per second. Another would describe a computer rate of one hundred twenty-five thousand operations per second and a third would brag of a computer working at the state of the art, i.e. two hundred twenty-five thousand operations per second. Then someone from IBM would say, "Our machine performs ten thousand operations per second for twenty-three hours out of twenty-four and we use the twenty-fourth hour for PM."

All the other people shutted up. Their high speed computers could not keeping working long enough to even approach the performance of the IBM machine.

W. A. Dickinson

I joined the Emporium Section of IRE (later the IEEE) when television was the prime topic for section meetings, conventions, and papers in *Proceedings*. (I later served a year as Chairman of the Emporium Section.)

Jerome Kurshan

I was fortunate in having two wonderful mentors in getting started. One was G. Ross Kilgore, a pioneer in magnetrons, who headed our vaccuum tube research group. The other was Edward W. Herold, a section head who subsequently advanced into a variety of technical management positions at RCA (with a brief interlude at Varian Associates).

Ed was truly inspiring: a very hard worker, a creative innovator, a clear thinker, a wide-ranging intellect and a warm human being. For them, IRE (predecessor of AIEE) was *the* professional society and I joined early on. (Ed stayed active in IEEE affairs throughout his career and continued to be involved in Life member activities in retirement.)

Their encouragement and support led me to present a joint technical paper on unclassified aspects of the magnetron research at the 1947 Annual Convention of the IRE in New York City. I became active in other IRE affairs as a member of several technical committees and worked my way up through the ranks to become Chairman of the Princeton Section (1954-55). It was good for my personal development and helped me develop relationships with my peers in other companies.

I remember one fringe benefit of these activities was the pleasure of occasionally communicating by phone with Emily Sirjane at IRE headquarters. I had never met her in person, but she was most charming and helpful.

Paul G. Cushman

Most of my previous reports, some of which conceivably might have been candidates for IEEE papers, were classified, as is the case for most work done in a department specializing in military equipment.

I served on a workshop group that was considering Standards in Control Systems and, for several years, reviewed and commented on Control System Papers that had been submitted for publication.

Most notably, I chaired one of the sessions at the national meeting in Seattle, Washington. I remember that session well because an airline strike was on and three of the presenters in my session did not make it. To partially fill in the time, I presented one of the papers myself. The co-chairman of that session was Mr. N. B. Nichols, the creator of "Nichol's Charts."

Hans K. Jenny

I joined the IRE as a student in Switzerland and have enjoyed my membership very much ever since. The IEEE, of course, is THE association which every engineer active in the electronics field should join and support. I have greatly profited from my membership by attending its conferences, presenting papers at its meetings and publishing articles in its periodicals. In 1966 I was elected a Fellow for "contributions to the development of micro-wave devices." I also participated in committees of JEDEC and EIA.

Chapter eleven

In retrospect

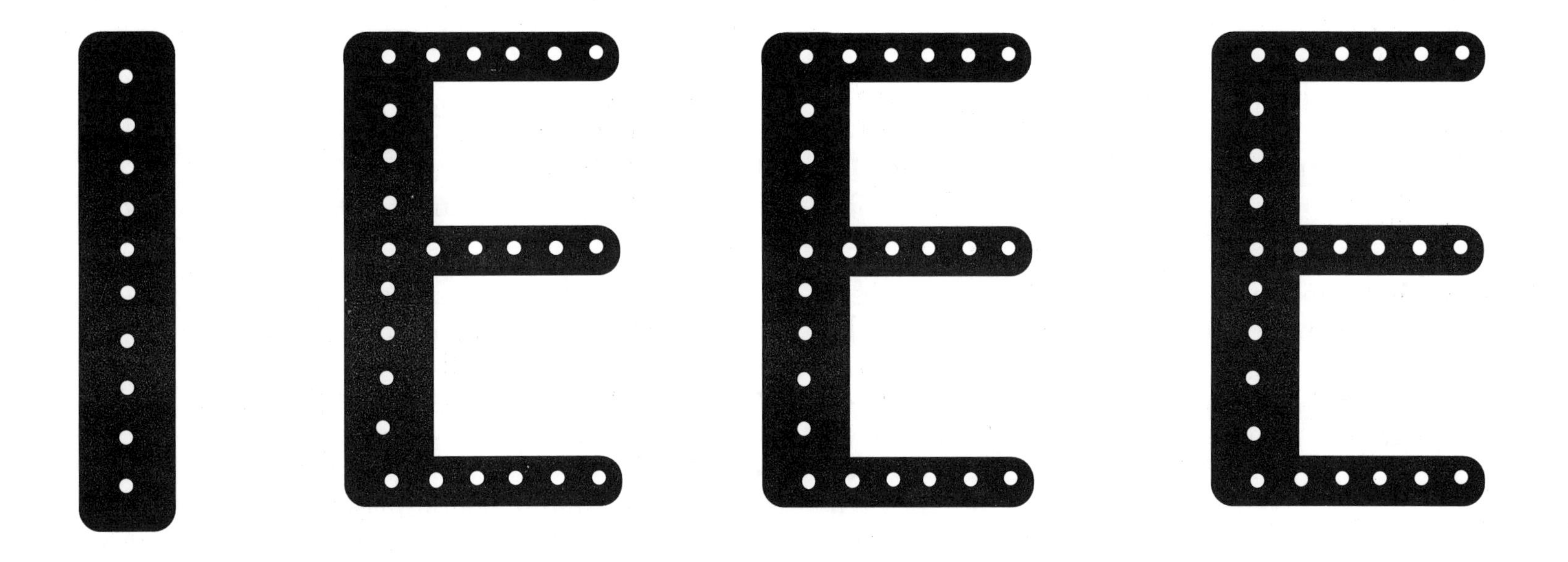

Paul D. Rogers

The vision of an engineering career had seized my imagination. In my fantasy, I was a creative engineering scientist; another Thomas A. Edison. My total fulfilment was through the satisfaction of my technically-orientated, creative urge. My total devotion was to my work, excluding all else—in particular humans, with their unpredictable and vulnerable emotions. Sequestered in my lonely laboratory, I turned out imaginative inventions as infinitum.

The real world of my subsequent engineering career did not exactly equate with my youthful fantasies; however, the basic direction had been established. My introverted thinking started me down a path isolated from the philosophical side of life and from my feelings and emotions. Only much later would I understand the underlying motivational factor in my decision to become an engineer—the protection of a sensitive ego.

My friends were chosen for their technical prowess, and all discussions thenceforth were technical jousts. My mind was continously cranking through formulas and strategies, leaving no time for illogical thoughts; no time for indulging in the deep mysteries of my non-technical self. My technical shell became impenetrable. In this realm, I excelled.

Considerations of my career and employer demanded priority status over personal and family matters. As my engineering career progressed, I became my own employer; owner of my own business. This new taskmaster always had priority. With the demands of my own business, I could now, more easily than ever, avoid those dreaded philosophical discussions with my wife and family.

However impossible it might seem, I went on to become a "successful" engineer without losing my wife and family, even to the time of retirement. Now what? What does a "retired" engineer do? No longer directly associated with the engineering industry, I was soon out of date in my technical knowledge so that any technical effort or discussion had lost its zest. All of my "technical discussion" friends had gone their own way, anyway. Our five children had left the nest to search for their own non-engineering success. That most of them chose non-engineering careers does not now seem surprising.

Without technical matters expanding to fill all available time, the days seem empty. I began searching for something that seemed to be missing. Just what was missing I didn't know, but it had something to do with life after engineering; or maybe just life, itself.

Back to the books, this time philosophical rather than technical. The sense of urgency to achieve was even greater than at the beginning of my engineering career. However, achievement was illusive. My standardized, long-proven technical study procedures didn't work. Now I was involved with things illogical. The answers were not given in the back of the book. Each study generated more questions than answers. My emotions were having difficulty coping, particularly given their torn and confused starting point.

But wait! Is not logical thought processes required for any advanced development to be successful? What better foundation for developing and expounding unique philosophical concepts than a discipined, logical mind? And what better logical thought training than an engineering career? It is high time that I apply my hard-won technical discipline to some really difficult non-technical problems. Is it possible that illogical introspective refelection can provide, for me, some logical guidance? Perhaps my engineering career can, after all, be *technically justified*.

Through diligent study and effort, I have now achieved a significant milestone in my philosophical understanding. I am now at the point, if I may use an electrical engineering analogy, of accepting Ohm's law. Although a rather basic beginning, I have already developed a feeling of satisfaction, coupled with a bit of wishful thinking; wishing that I found more appropriate guidance at the time of my career choice.

Simpson Linke

In retrospect, I derived great satisfaction from my varied forty year career at Cornell University and associated organizations. I engaged in exciting activities, encountered challenging problems, and had the pleasure of meeting and working closely with many talented individuals. I like to think that my contributions to various fields have been of some value. For the most part, I was given free rein to choose my research areas and to conduct my teaching activities as I saw fit.

My greatest satisfaction, of course, stems from my belief that I had some influence on the hundreds of students I had contact with at Cornell. On occasion, there is evidence that I was able to transmit my enthusiasm for my field to a substantial number of students, and that I even challenged a few to greater individual efforts. As a teacher, I can receive no greater accolade.

Garabed Hovhanesian

I enjoyed all of my jobs at General Electric (GE) though I did not enjoy getting caught in the political turmoil. I owe a great many people in and out of the company for my successes. I learned that I did not want to work for a manager I could not respect as a person and as a professional; I expected the same sentiment from my manager with respect to me .

I cite a comment made by Admiral Grace Hopper, "We manage things, we lead people." Management in, as well as outside of, GE in those days had a lot to learn. One must want to do what one wants to do; simply taking a job to fill a resume can make a person very unhappy. Whether you are tackling the laws of nature or whether you are tackling the laws of human behavior, one must enjoy what one does. Only then can you look back on your life and say, "I'd do it all over again."

C. R. Schmidt

An engineer's first goal is to design a product which is not only useful but also makes money. As far as permanence is concerned, no matter how original an engineer's design may be it will never last forever. The inexorable progress of technology sees to it that engineering designs will find it hard to be in continuous production for thirty to forty years. So the practice of engineering may not achieve permanent results, but it certainly is an absorbing adventure.

The scenario today begins with the marketer (usually with management's blessing) presenting the engineering group with the customer's wish list for the product and suggesting the direction the product design should take. The engineering segment, after commenting adversely on the barrenness of marketing's design suggestion, then presents a menu of product benefits versus estimated cost per benefit.

The marketing people usually don't know exactly what the customer wants. The problem of obtaining the customer's true need involves making him or her understand that little-used features will add to the product's cost without making it substantially more useful. Once the customer's thinking is refined to consider an optimum design, his or her true requirements will emerge.

A design engineer who can think in terms of hardware required to provide these benefits can be very helpful at this stage. In 1986, it was suggested by an engineering vice president at General Electric that a design engineer accompany a sales or marketing person to call on the customer for the purpose of asking the right questions. Successful designs result when the product requirements are optimized. This is best achieved when one person makes the tradeoffs.

This is because the committee approach is doomed to failure from the start. The trouble with design committees are that the product design becomes a compromise in which each participant tolerates the others' ideas, so that all pet ideas are included somehow. This is far from an optimization process.

The complete design should be made by one knowledgeable engineer. If his or her design is not acceptable, he or she is the wrong person for the job and another engineer should be selected. The design engineer must not retreat to an ivory tower in the design process. On the contrary, he or she should try to include all the suggestions presented and through discussion of individual requirements with others clarify all the possibilities in his or her own mind. The process should include the maximum number of significant benefits with the minimum of hardware and cost. It can be a very engaging process especially when it's on the right track.

In this process, the engineer can have no clever or cute pet ideas that he or she is stuck on. The optimum design must include the best, most suitable ideas, whether they're the design engineer's or somebody else's. The end product must be considered and visualized completely in the designer's imagination before the final specifications and sketches are made. In some cases, working prototypes may have to be made to prove feasibility or suitability of some functions. Once these determinations have been made no additional detail designing should be undertaken until the complete product design has been accepted.

If there are competing approaches to the product design, each should be considered thoroughly and objectively. Only then can the selection from the alternative designs be made. The best design is then presented to sales, marketing and management for acceptance. If the job has been done properly the prospect for acceptance will be quite bright.

The constraints of meeting the target design specifications and target price goals are commonly understood by project engineers. The constraints of project expense and date of completion of the product design are grudgingly recognized or not recognized at all by many project engineers. This is particularly true if the project expense budget is low and the time to completion is short.

Engineering design time is a company investment and as such cannot be squandered. So if the budget is very tight, the engineer may be obliged to restrict himself or herself to a choice of known technology or possibly to just an extension of the company's existing product line. No examination of emerging technology may be possible. This obviously will not result in one of the company's future breadwinners, but a return on investment has to be anticipated at some point. However, if company management regularly short budgets its products, it will be condemning the company to new product offerings not keeping up with the new technology. Consequently, business volume will decline.

Management must recognize the talented engineer and assign him or her to short budget projects. Chances are, the talented engineer will accept the challenge and produce a beneficial result. However, there is no way even the most talented engineer can overcome a severely short time and money budget. Management is responsible for this phase of the project. There is no free lunch in the free enterprise system. If management cannot see a substantial return on its investment with a new product consistent with its established quality level, the project should never begin. An engineering bail out is a long shot maneuver undertaken only by start-up companies and companies about to go out of business.

Regardless of the circumstances, the engineer will make the compromises needed to meet the price and performance specifications in the time and money limits allotted for the project. This type of performance is what has made engineers worth the money they are paid today and, indirectly, is what has brought about their relatively high starting salaries.

In many successful projects, I have felt, after the money starts to be made, that I had only a small part in the outcome and that all the others had done much more than I. Perhaps this feeling is the result of another job related attitude that I have.

I listen to the members of my team, whether they are technicians, design draftsmen, or purchasing agents. In discussing the various aspects of the project with them, I encourage them to make suggestions and observations. You encourage people when you don't harshly criticize their offerings and when you entertain their suggestions to a conclusion.

Although I was not the project engineer at the time, the following example illustrates this point. In the late sixties, we found that the electronic parts of our instrument designs were much lower than the cost of the instrument enclosures—due mainly to lower semiconductor and power supply costs.

Together, with the corporate industrial design group, we cast about for alternatives to the sheet metal fabricated construction we were using at the time. We explored the possibility of extruded aluminum side sections with mitered "V" grooves as fold points to make up the box together with a sheet aluminum bottom captured by extruded grooves. This was attractive because extrusion die costs were low. We even explored sand castings, one of which won a design institute prize at a national instrument show.

Our problem was that our new design sales quantity seldom exceeded one hundred units a year. So we were trapped between high fabrication costs and high tooling costs. In the course of our discussions with the corporate industrial designer, he mentioned that "if the annual instrument volume is greater than one thousand a year, the California designers consider die castings." Those were magic words.

At our operation, we had an ancient line of slide back conductivity instruments with fabricated cases that were long overdue for a redesign. The combined annual volume at that time for these instruments was eighteen hundred pieces. The proposal was made that we make a universal case design that could be used for both the old slide back and new meter type instruments.

The old enclosure costs were about twenty-eight dollars a piece, the new die cast case cost would be about two dollars. This would bring the enclosure cost of new designs in line with the lower electronics costs. The return on investment in the die casting was several hundred percent in the next five years well above the fifty percent required by corporate accounting.

This design process—in which a part to be used in new production is also used to improve the profitability of old production—is called contextual engineering. It is just one way to restore yesterday's breadwinner products to good profitability. Listening for the magic words is a habit that can produce very beneficial results.

George V. Jacoby

(M '51; SM '58; F '81) I received the Diploma Ing. degree in electrical engineering from the Technical University of Budapest, Hungary, in 1941.

In the United States, I worked with Honeywell on research and development of industrial instruments, optimizing control methods and servo mechanisms. In 1958, I joined RCA, and worked on various phases of magnetic recording, servo systems and new equalization methods for digital tape recorders and disk files. In 1965-66, I invented and developed into a new product, the Delay Modulation code, with which I doubled reliably the linear bit density of the previously standard Manchester code. Five years later, the same technique became generally accepted by the digital recording industry, as the MFM code.

In 1971, I joined Sperry Univac working as Manager, Advanced Recording Techniques, and later as Senior Professional Consultant. In the mid-1970's, I invented the 3PM code which I provided a reliable fifty percent density increase over the MFM code in a new high density disk file. For this invention, I received the ISS/Sperry Univac Outstanding Contributor's Award in 1978. I hold twenty patents with several pending.

Charles T. Morrow

An engineer is a leader of men. There may be more individual engineering consultants than ever before, but today the engineer is more typically part of a team—sometimes an extremely large one. He should be proud of his subordinates and never feel the need to compete with them. One measure of his professionalism is the pride he takes in his technicians. The skills of a technician are important. Professor Percy W. Bridgeman of Harvard qualified for the Nobel Prize in part by being a superior technician as well as a superior scientist.

A common saying is, "Power corrupts. Absolute power corrupts absolutely." Yes, power can be used irresponsibly or vindictively, but a power vacuum can be a prelude to the abuse of power. There was a power vacuum before the rise of Stalin and before the rise of Hitler. Power should be respected and used wisely.

Another popular saying is, "There has never been an arms race that did not lead to war." Perhaps so, but there have not been many "arms races," as we think of the term now, to test it. I would certainly not favor one if it could safely be avoided. It is worth reflecting, though, that the lack of an arms race contributed to the beginning of World War II. England and France were not prepared to confront Hitler before world war became inevitable.

We have also taken more trips overseas, east, west and south. We are beginning to understand how the world is put together and what its people are like. Two equally perceptive people may travel abroad. One may report that people are all the same. The other may report that they are entirely different. They may both be right, but they are not talking about the same thing. The first is talking about fundamental motivations. The second is talking about how these motivations are expressed as customs, depending upon environment, history and tradition. Here is a paradox: If one cannot tolerate the differences, one cannot see the similarities.

I have not reached the exact top of the administrative mountain or the technical mountain, but I have had experience with both. In a mountain climb, some of the best views are part way up.

Herbert Matare

A short story of my professional activities:

- The drive to be at the forefront of technology often means that valuable thoughts and projects remain dormant for a long time. Not the best and most advanced ideas can take hold and result in new activities if the time is not right for their applications.
- Much of my work was of a more pioneering nature and always somewhat ahead of industrial exploitation. I could have continued only in a few very large industrial research laboratories or a few universities. Thus I started my own consulting operation, with the known problems of finding suitable environments for advanced work.
- It is the dilemma of those doing advanced work, to have to get down to more mundane, market-oriented work. Otherwise, they must adapt to the difficult situation of pursuing their inclinations to do advanced work with the attendant tiring efforts of finding appropriate funding.

Enrico Levi

The wisdom that is supposed to come with age entitles me to pontificate, so I will now offer some advice to the young.

First a few do's. Do what you like best, if you have a choice. If you don't, get deeply involved in what you are asked to do and you will end up liking it. Make your work your hobby. Do engineering, if you are technically inclined. It is a very rewarding profession, even with salaries what they are today.

Get a broad-based undergraduate

education. Computer science notwithstanding, take as much math as you can. This opens the door to an understanding of the physical world and its wonders. Start your career with good hands-on, field experience. Stay near a good university and persevere with your education. Keep yourself well informed by joining the IEEE and other professional societies and by attending their meetings.

Now a few don'ts. Once you have made a decision, never look back and ask: Was that the right thing to do? Don't rush into specialization at the beginning of your career, and into management later. Above all, never despair!

Frank T. Luff

I had no mentor and no particular role model. I had some good bosses and some bad. Most seemed too much involved in empire building and work place politics. Peer relationships were sometimes good and sometimes bad. The demands of my job and family were always too great to permit participation in IEEE or similar activities.

There were periods (sometimes years) when it seemed the whole world was trying to prevent me from realizing my career potential. In the end, I feel that I made a difference in my field of technology. I worked hard at it. I feel that I came a long way from where I started. However, I could have gone further if I had been able to get a better basic education earlier in life. There were ethical dilemmas during my career. But these were not a big factor. I prided myself in high and unbending ethical standards. Engineering principles are another matter. By this I mean, many times an engineer has to accept changes in design and/or cost estimates ordered by someone else or find another job.

Engineering principles are similar to the academic freedom of a teacher. Both are of great concern to me. I believe this is a problem not only in the engineering and academic fields but throughout the work place in our society.

Thelma Estrin

Both my husband and I are IEEE Life Fellows, but we are in different fields of work. From the beginning, our marriage of fifty years has been a partnership, with each of us possessing different attributes. Despite what our fellow students thought, however, we never studied together because our work styles were so different.

The present increase of women in engineering will not only fill the gap for needed engineers, but will also bring increased diversity and creativity to our field. Here's to future women Life members of IEEE.

Leo Berberich

Looking back, I feel that at age forty-three, without an extensive management background or, better still, an MBA, I probably should have stayed in reasearch and development. My career never did advance beyond middle management.

While the liaison and international positions offered some very attractive and interesting compensations, none were as intellectually stimulating as research and development. I enjoyed the work that resulted in new patents for Westinghouse and the publication of papers the most. If I had remained in the lab, I would undoubtedly have achieved a more impressive international reputation in my field and collected more significant awards.

Self-satisfaction just wasn't present in the latter part my career. However, I have few regrets about the path taken. I tremendously enjoyed the world-wide travel, visiting sights that most people never have the opportunity to see.

I feel that the high point was the years I spent in research and development The low point was the months after finding my operation in Geneva was being shut down. I recognized too late in life that large corporations like Westinghouse are not run by the Ph.D's but by the MBA's.

Rowland Medler

So, in 1980, after twenty-two years with WUFT, four assignments and relinquishments as "Engineering Manager," with never being given the authority to manage anything, I became eligible for early retirement. I jumped at the life saving opportunity. Under the same circumstances, I'd probably do it again. The past eight years have been the best of my life.

Our suggested agendas for these treatises doesn't include a request for advice to our younger followers. So, in my typical contrary manner I'll give advice anyhow. It's simple: Get that piece of paper! A college degree, in the eyes of a prospective employer, is worth a lifetime of hard experience. You'll stumble into the experience as you muddle along. But it's doubtful you'll stumble into the degree.

In my case, the difference, over the years, has probably amounted to half a million dollars. You may, like me, end up with a clear conscience and an internal feeling of great accomplishment; however, it won't show to anybody except those with persistence enough to read reports like this one.

Robert McLane

My greatest accomplishment was being part of the team which gave pilot Bob McGregor such a warm smile of appreciative confidence after his first flight with our automatic "flare-out" computer, blind landing system. The U.S. Air Force sponsor judged it to be the finest system they had funded and flown. It made many friends in the Air Force, but the development of microwave landing systems of that time, availability of test-bed aircraft and other scheduling obligations of the F-94C demonstration vehicle signaled the end of that great program.

My worst failure was the aforementioned bout with office politics in the human factors proposal debacle. As stated, I would have documented my no-bid judgment, had I been aware that spoken words in meetings quickly fade in management memory.

Retirement time arrived when my best friend, my wife, advised that I request it! I didn't realize that I was ready, but she saw it clearly.

Age discrimination was keenly felt in a field support assignment when I was judged "overqualified," and was subsequently returned to home office for lay-off or triple demotion, my choice.

Yes, I am very glad that I worked and lived my professional life mainly as an electrical engineer. I am, also, delighted with my second career choice as a realtor. I now can merge the precision and care of detail with the emotional world of home value assessment!

Melvin Manning

The future for elecrical engineers is very good, but many more power engineers should be educated in universities. Faculty is at fault many times, as they have no industrial experience and place too much emphasis on Ph.D. degrees. Computers are getting too much attention at times. The fundamentals of engineering are neglected. Ethical problems do exist and cost is always the bottom line.

John J. Dougherty

Most of my fellow naval engineers, contemporary, junior and senior, helped me learn that leadership can be reduced to three different kinds of knowledge; two important ones being knowlege of one's (technical) "stuff" and knowledge of one's associates, bosses and juniors. These, however, cannot be learned satisfactorily until the knowledge of one's self— the most difficult knowlege of them all— is well on its way to being understood. Without this knowledge, one can smash oneself on the rocks of compulsive perfection.

Lester E. Haining

From horse-drawn transportation when I was a child to jet-borne travels to far away continents today; from telephones of the "hello, central" type to dialing an individual phone to any of many foreign countries from my back yard; from active components like the 201A tubes which were my first introduction to the field to elaborate computers on chips; from radios like the first one in my family which was a TRF type using A, B, and C batteries to television from all which I watched a man walking on the moon—what a time to be alive and be a part of the action! We have seen greater changes and advances that can make life better for the human race than occurred in many centuries before us. Will any generation to come experience anything more interesting?

C. Richard Ellis

I am happy that I chose the engineering profession. In the small mid-west town where I was born in 1920, a large number of the homes did not have indoor plumbing or city water. A newborn then was almost certain to have measles, chicken-pox, mumps and whooping cough, during its childhood. There was also a good possibility of catching a more serious disease: small pox, scarlet fever or polio.

There were a few battery operated radios and wind up phonographs. Nearly every in-town house had a connection to the local electric plant and had a few light bulbs hanging from the ceilings. The farm homes had kerosene lamps. Only a few of the well-to-do had cars. No one had air conditioning or a clothes dryer. Ice was the only refrigeration and there were few powered washing machines.

In the historically short seventy years since, there have been drastic changes in all these areas for the better. Much of the improvement has been due to engineering efforts.

Engineers designed the water and sewer systems and the machinery and techniques to allow mass production of pipes and pumps. They designed bio-chemical machinery to allow the mass production of serum for vaccinations. They conceived and developed the myriad of electronic devices allowing mass production of radios, TVs, sound and video recording and reproducing devices. They developed electronic computers and helped make the computer industry a giant. They developed medical diagnostic and treatment equipment allowing previously impossible repairs to the body. And engineers designed and helped construct the huge power plants and distribution systems that energize it all. Best of all for the engineers—most of them had fun doing it. I did.

Edgar C. Gentle, Jr.

What did we forget? The thing I expect we might have overlooked in documenting all of this is the need for the engineer, or the professional person who moves along into management, to make room for recreation and the enjoyment and responsibility of his or her family. This can take a lot of balancing if the job requires travel.

A person is fortunate when he or she finds someone to spend the rest of his or her life with. I've been most fortunate because Jeanne, my wife, is a very understanding, warm and compassionate person who makes friends easily and is respected and loved by our family (myself, three daughters, three sons and several grandchildren). Jeanne holds no grudges and we try not to dwell on "looking back." We have lived in many places and enjoyed each location.

In summary, I've been very fortunate to have had my professional training and the help from all the people with whom I have worked and who assisted and advanced me. I do thank God for the chance to live this career as I have done.

W. A. Dickinson

My work experience was unusual. I stayed with one organization and worked on one product line for the entire time.

Meetings are often important, but I sat through too many which I considered a waste of time. A meeting at which decisions are to be made requires a string chairman, freedom of honest participation by those involved, and written record of conclusions. I believed in putting things in writing; nobody put out more one-page memos than I; they prevented misunderstandings. I also believed in good, complete manufacturing specifications, and in adherence to them.

The designer starts with something that exists, then modifies it to improve it, or makes a new product. He or she must understand the principles behind the operation of the product and work with, not against them. The designer should observe how a thing "wants to work" and make it that way. The more he or she knows about materials, manufacturing processes, and the intended application, the better the designs can be. The product should be as simple as possible, reduced to its essentials, and made the easiest way.

Paul G. Cushman

When reviewing and evaluating a career it is sometimes depressing to remember the frustrations, failures and missed opportunities. It may be helpful for us to bear in mind that only a relatively few can be "rich and famous." One criterion of "success" we can use is the enjoyment our jobs gave us, and our feeling of accomplishment in doing a good job, at whatever level in the "hierachy" we happened to be.

I fully enjoyed the cooperative tasks with my peers, during which talents and technical contributions were mutually recognized and appreciated. Early retirement was never a consideration for me. Health problems in my family made it necessary to retire soon after the age of sixty-five. Otherwise, I would have continued at the job for several more years.

Franklin Offner

Boss, mentor, and peer relationships. My only "boss" in industry was Sid Shure. He helped me greatly by refusing to give me a ten dollar a week raise.

In education, I was fortunate to have an excellent science teacher in high school, as well as good courses in drafting and manual arts. (In those days, using your hands as well as your brains was not considered demeaning.) These high school courses served me well later in life. At Cornell, my course in electricity and magnetism by C. C. Murdoch served as the basis for my work in electronics. At CIT, working with Linus Pauling and his colleagues made me aware of what constituted real intellectual greatness, as did my later contacts with Carl Eckert at the University of Chicago. In contrast, another professor at the University of Chicago exerted a strong negative influence on me, as well as others working under him.

Shohei Takada

Almost finishing my career life, I have become aware that it is important for Japanese engineering to connect the fundamental research and practical engineering more tightly. So, fifteen years ago, I established a tiny company (LINK Laboratory Incorporated) to link the two fields.

Samuel Sensiper

In reviewing my career, I have been fortunate, perhaps, that I never faced any major ethical dilemmas (which I know other engineers have faced). In the few instances where it appeared there were minor ethical problems, these usually turned out to be resolvable by confrontation and discussion. I have no regrets and I do not believe I have harmed anyone. As to whether I have been harmed, I have always felt that I was responsible for my actions and should take the blame or be given the credit for what I did. If I have received more or less of the former or latter, then again the fault is mine and not others. If I have any regrets, it is that I was not more entrepreneurially oriented in my twenties and thirties when the penalties for failure would not have been so large. And the chances for success would have been larger.

J. Rennie Whitehead

For productive innovation, I found that creativity tends to be stifled in any line organization, whether it be industry, university faculty or government department. The reason is simple: the line departments are restricted to pursuing the primary aims of the organization, which, for industries, is survival and profit; for universities, the production of educated people, and, for government, the line functions of policy and regulation.

Only by decoupling R & D, from the day-to-day line functions, can the essential full-time group of viable size be assembled and have the freedom to look at the long term without the distraction of immediate administrative problems. I found that it did not matter much whether such a quasi-independent establishment was affiliated to industry, university or government. The appropriate degree of decoupling from the line organization was what counted. As a consultant, I find that retirement has come on imperceptibly. It merely means taking on fewer commitments each year and finding more time for those nonpaying, personal projects that always seemed to be put off from year to year in the past.

Alfred J. Siegmeth

I envisioned already, around 1955, that earth satellites would provide continuous surveillance of the surface of our earth and would detect military activities by any nation. International news acquired and dispersed via media elements, such as radio, TV and newspapers, would inhibit monarchs, dictators and key political functionaries to build up their own, super empires by defrauding public funds and national treasures.

My studies of humankind's history have shown that all wars or even skirmishes were initiated by insufficient, untrue and twisted information exchanges. As a regular rule, monarchs declared wars without the knowledge and approval of their subjects and their military. The world wide development of the expanding telecommunications systems of the INFORMATION AGE will be the real guardians of PEACE. The World's harmony must be maintained and controlled by well planned and organized information exchanges, negotiations and treaties. To enhance the life standards of undeveloped nations, they must participate in an intensive knowledge exchange methods.

Hugh J. Cameron

A dominant aspect of electromagnetic engineering is the continuous reversal of the flow of energy. The reverse is true of human communication. Feedback is engineered into many of the devices we create, but is omitted from both casual and monumental personal relations.

I have written and delivered several training courses. I quickly learned that I could not transmit even elementary concepts, if I occupied the rostrum more than 25% of the time. I forced myself to be a nonentity in the classroom. When the pupils discussed the concepts with each other, they acquired a visualization of the factors involved. At first I had serious objections, but the eventual results were greatly accelerated learning.

I recently audited a sophomore electrical engineering class with my grandson at U.C., San Diego. The subject was the electrical charge on two spheres. I know the subject, but I could not conceive that one of the hundred or more students acquired anything from the one hour lecture.

R. H. Eberstadt

This is the point for which I feel proud to have had the opportunity to have lived and worked during this period of time. Having played a part in the industrialization process of my country, and also the geographical region where I have been active for about twenty-five years, has given me the chance to help develop tens of thousands of human beings, to permit them to become active in creating something for themselves and their families. Some I have personally known and encouraged. Therefore, I have been able to witness the multiplier effect they have produced.

Melvin L. Manning

Future for Electrical Engineers

Very good, but many more power engineers should be educated in universities. The faculty often have no industrial experience and place too much emphasis on Ph.D. degrees. Computers, at times, get too much attention and the fundamentals of engineering are neglected.

Ethical problems exist, but cost is always the bottom line.

Questions and answers

Jay Patchell

Q: *Did you stay within the system throughout your career or did you branch out on your own or with others?*

A: IBM was a dead end in engineering so I quit. At Raytheon, I entered power conversion. The specialty was very low noise wide bandwidth amplifiers to tune BWO's (backward wave oscillators).

Q: *What were the little frictions, miscommunications or directives between you and your colleagues that would interfere with work?*

A: I was mostly independent. Other engineers couldn't understand my emphasis on detail.

Q: *What were the attitudes of people in other parts of the system (e.g. the marketing department in a corporation) toward your area? How did their attitudes help or hinder your work and career?*

A: My ability to design to the limits of available components and keeping abreast of component developments helped my career. My last decade at work I did trouble shooting when systems were in trouble from power converters, circuits and microwaves.

Q: *How did world events, such as the Depression and World War II, affect you personally and professionally?*

A: WWII changed my career from building trades to electronics. My career growth was held in check by a manager who didn't like me.

Q: *What major and minor ethical dilemmas did you encounter at various times and how did you deal with the situations?*

A: I challenged my manager's ethics whenever I thought he was trying to obscure the facts.

Q: *When did you join IEEE (AIEE / IRE)? Did the organization live up to your expectations of it?*
A: I joined both in 1948. Twice in my career I found articles that helped me solve difficult problems.

Q: *Looking back on your career, what do you see as your greatest accomplishment?*
A: At Raytheon, I was assigned a task others wouldn't touch for fear of failure. The Microwave Department manager praised my work, however, my own manager belittled my work.

Q: *What else would you like us to know about you and your work?*
A: I started an electronic music service about ten years before retirement. I had a good business operation going upon retirement. It has been very active ever since, because I am the only "house call" service man in a one hundred square mile area. My technical experience sure helps even now.

I would like to advise other engineers who have knowledge of computer circuits and digital devices that many opportunities requiring their knowledge exist. The best is servicing industrial electronically controlled machines and robots, music instruments and other electronic devices from old tube radios and such. I am not busy one hundred percent of my time. Thus, I now have time leftover to pursue my model railroading hobby in which I design gadgets with electronic control.

Alvin J. Markwardt

Q: *How did you deal with the introduction of new technologies?*
A: I mastered it.

Q: *What were the new technologies?*
A: Computers.

Q: *Did you feel threatened by them?*
A: No.

Q: *Did your company provide assistance in keeping on top of new developments?*
A: Yes.

Q: *At what age did you feel the most roadblocks to your career?*
A: Thirty-six. This was in 1956 when the large turbine generator technology advance stopped in the U.K. I moved to Canada and then to the U.S. to progress in this field.

Q: *When did you join the IEEE (AIEE / IRE)?*
A: I joined the IEE in 1938 and the AIEE in 1950. This was to obtain technical information from journals in my field.

Q: *When retirement time arrived, were you ready for it?*
A: Yes.

Q: *Did you ever feel that you were a victim of age discrimination?*
A: No.

Q: *Who were the most important people in your career and who were the least encouraging?*

A: The most important was Dr. Lee Kilgore. The least encouraging, money controllers.

Q: *Are you glad you lived your life as an EE?*
A: Yes.

Darrel E. Moll

Q: *Why did you become an engineer?*
A: I didn't have enough money to become a doctor.

Q: *When did you first start thinking about engineering as a career?*
A: K. P. L. offered me a job at Manhattan, Kansas where I could go to Kansas State University.

Q: *Who were your role models?*
A: W.S. Wick, K.P.L. Superintendent at Hutchinson, Kansas.

Q: *Did you have a mentor?*
A: Yes, Miss Helen Moore at Hutchinson Junior College and later at Kansas State University—the Dean of Women, Math.

Q: *What was your first job?*
A: I was a draftsman at K.P.L.

Q: *How did you get it?*
A: Through a recommendation from Hutch Junior College.

Q: *What were your responsibilities on your first job as an engineer?*
A: I worked at a small power plant on transmission lines where I gave job assignments and materials to line foremen.

Q: *What technical area did you specialize in?*
A: Power engineering.

Q: *Why did you choose that specialty?*
A: It was my first job.

Q: *Did your career in this area progress smoothly or did you move around within various technical specialties and job descriptions?*
A: After the war, I changed to telephone toll transmission testing methods and test set design. My next step was mechanical engineering on gas turbines, then testing of purchased items.

Q: *Did you stay at the drafting board or move into management?*
A: I was the Power Department Chief for five years.

Q: *What jobs, projects, special assignments met your work expectations, exceeded them or came up short?*
A: Test set design was enjoyable.

Q: *How much freedom were you allowed in producing results?*
A: I had very little supervision.

Q: *What were the little frictions, miscommunications or directives that would interfere with your work?*
A: I had two "bad" supervisors.

Q: *Would you have changed the organizational structure you operated in to increase job satisfaction and results?*
A: In raise and appraisal conferences, I would allow equal time to each supervisor for his men and not just the loudest mouth and the fastest tongue taking over the conference.

Q: *What were the new technologies you had to deal with?*
A: Transistor technology.

Q: *Did you feel threatened by the new technologies?*
A: Not at first, but after five years as a supervisor, there was too much progress.

Q: *How did world events, such as the Depression and World War II affect you personally?*
A: I lost five years.

Q: *How do you view your career growth now and how did you feel about it then?*
A: Not a problem. I'm retired now and I am enjoying life.

Q: *At what age did you feel the most roadblocks to your career?*
A: After age 52. Younger engineers were handed the better projects.

Q: *What major and minor ethical dilemmas did you encounter at various times?*
A: Suppliers offered too much entertainment.

Q: *Were you an observer or active participant and how did you deal with the situations?*
A: We just said we couldn't accept it.

Q: *Did the IEEE (AIEE/IRE) live up to your expectations when joining?*
A: I enjoy the lectures.

Q: *Did you join other professional associations?*
A: No.

Q: *Looking back on your career, what do you see as your greatest accomplishment?*
A: Test sets designed for installers were demanded by telephone companies for maintenance.

Q: *Looking back on your career, what was your worst failure?*
A: I couldn't get a fair share of the raises for my men.

Q: *When retirement time arrived, were you ready for it?*
A: Yes.

Q: *Are you glad you lived your life as an EE?*
A: Yes.

Legacies INDEX

Legacies INDEX

Samuel E. Spittle

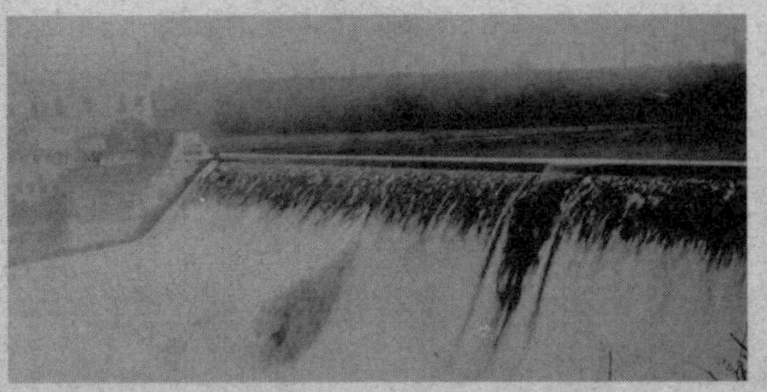

All the pictures were taken on a field trip of the Oregon State EE senior class in the spring of 1928, by some class members. All of the plants shown were operated by Portland Electric Power Co. (now the Portland General Electric Co.), and were located within 50 miles of Portland, Oregon. As far as I can remember, the River Mill (upper right), Cazadero (upper left) and Bull Run (lower left) plants were located on the Clackamas River, which runs into Willamette above Portland.

There where at least three IEEE Life members, who were members of my class and on the same field trip. Their names are: Charles Savage, Richard Setterstrom and W. A. McMorris.

The River Mill plant had a head of approximately 80 feet, the water being supplied from a dam on the river. The dam was of somewhat unusual construction, being hollow, like the roof of a house, with a walkway through it.

The Cazadero plant had a head of 125 feet. I don't know whether the water was supplied from the Clackamas River or a tributary.

The Bull Run plant was located on the bank of the Clackamas, but was suppled by water from Bull Run, a stream which also was the main water supply for the city of Portland.

Samuel E. Spittle

All the pictures were taken on a field trip of the Oregon State EE senior class in the spring of 1928, by some class members. All of the plants shown were operated by Portland Electric Power Co. (now the Portland General Electric Co.), and were located within 50 miles of Portland, Oregon. As far as I can remember, the River Mill (upper right), Cazadero (upper left) and Bull Run (lower left) plants were located on the Clackamas River, which runs into Willamette above Portland.

There where at least three IEEE Life members, who were members of my class and on the same field trip. Their names are: Charles Savage, Richard Setterstrom and W. A. McMorris.

The River Mill plant had a head of approximately 80 feet, the water being supplied from a dam on the river. The dam was of somewhat unusual construction, being hollow, like the roof of a house, with a walkway through it.

The Cazadero plant had a head of 125 feet. I don't know whether the water was supplied from the Clackamas River or a tributary.

The Bull Run plant was located on the bank of the Clackamas, but was suppled by water from Bull Run, a stream which also was the main water supply for the city of Portland.